Ricardo Luiz Nogueira de Souza

Restoring the Atlantic Forest:

Ricardo Luiz Nogueira de Souza

Restoring the Atlantic Forest:

potentialities, fragilities and environmental conflicts in the Mário Xavier National Forest/RJ

ScienciaScripts

Imprint
Any brand names and product names mentioned in this book are subject to trademark, brand or patent protection and are trademarks or registered trademarks of their respective holders. The use of brand names, product names, common names, trade names, product descriptions etc. even without a particular marking in this work is in no way to be construed to mean that such names may be regarded as unrestricted in respect of trademark and brand protection legislation and could thus be used by anyone.

Cover image: www.ingimage.com

This book is a translation from the original published under ISBN 978-620-2-17538-8.

Publisher:
Sciencia Scripts
is a trademark of
Dodo Books Indian Ocean Ltd. and OmniScriptum S.R.L publishing group

120 High Road, East Finchley, London, N2 9ED, United Kingdom
Str. Armeneasca 28/1, office 1, Chisinau MD-2012, Republic of Moldova, Europe
Printed at: see last page
ISBN: 978-620-7-97290-6

Table of contents:

I would like to thank my grandfather Sizenando Nogueira and my grandmother Leocádia Sbruzzi for their good example for as long as I live inspired courage and knowledge.

ACKNOWLEDGMENTS

To my advisor, Prof. Heitor Soares de Farias, a special thank you for everything. For the support and trust placed in me. For the true "north" of his guidance, for his constant encouragement and support, for his dedication and partnership and for all the knowledge he passed on during the course of this research. The true source of my reflections and inspirations and the person most responsible for my decision to enter academic life. To you, my sincere thanks and eternal gratitude.

To Prof. Leandro Dias de Oliveira of the PPGEO, for his clarifications and always constructive criticism in the development of this research.

To Prof. André Santos Rocha of the PPGEO, who was part of the qualification panel, for all his observations and contributions to this dissertation.

To Prof. Francisco Carlos de Francisco (*in memoriam*), for being the great creator of this moment in my life.

To my colleagues at PPGEO/UFRRJ and IM, for the countless learning opportunities and excellent socializing.

To my work colleagues at ICMBio for their constant support and encouragement in the final and definitive moments of writing this dissertation. In particular, Andrea Ribeiro, for her support and understanding, Paulo Motta for his encouragement and the FLONA MX employees, Abílio Gonçalves, Roosevelt Antunes, Roberto Matias, Carlos Eduardo, Romildo Pereira, Jair Costa, Gilvan Rangel and all the employees of the Companhia Brasileira de Trens Urbanos (CBTU), for their endless conversations of encouragement and strength.

To my beloved family. To my parents, Emil and Nair, who have always believed in me, who have been there for me every step of the way and who have always given me their unconditional support. To my brothers for their constant support and friendship. To my nephews Fernanda and Thiago for their affection and for cheering me on so that everything would work out.

In particular, I would like to thank my wife Lívia Letícia and my daughter Marianna Carvalho for their understanding and unconditional support throughout the time I spent doing this research.

To God above everything and everyone, the source of infinite light and goodness.

> "When someone evolves, everything around them evolves."
>
> Paulo Coelho

SUMMARY

SOUZA, Ricardo L N. **Restored Atlantic Forest: potentialities, fragilities and environmental conflicts in the Mário Xavier National Forest, Seropédica/RJ**. 2017. 90 p. Dissertation (Master's Degree in Geography) Agronomy Institute, Multidisciplinary Institute, Federal Rural University of Rio de Janeiro, Seropédica, RJ, 2017.

This work deals with the geo-environmental analysis of the Mário Xavier National Forest, in the municipality of Seropédica - RJ, using geoprocessing as a tool. It aims to provide a systematized understanding of proposals for ecological-environmental zoning, combined with the growing interest in the conservation potential of the Conservation Unit. The discussion aims to problematize the maintenance of this protected area insofar as it still suffers from various environmental conflicts, under the effects of anthropic action, related to the expansion of urban areas, as well as the recently created BR-493 highway, which sectioned off FLONA MX. Furthermore, it can be seen that the recognition of environmental conflicts is the basis of the proposal for environmental zoning, which justifies its maintenance as a conservation unit. This proposal works primarily to locate the potential, such as endemic species at risk of extinction (fish and radon) and vegetation (native and exotic), and the weaknesses. In order to understand the subject, a legal framework was briefly presented, focusing on the concept of Conservation Units, and the documents of the unit studied were analyzed, as well as the results of the environmental zoning proposal. To this end, maps of the potentialities were generated (land use and land cover map). In order to analyze conflicts, a mapping exercise was carried out in which some of the peculiar characteristics of FLONA MX are presented, such as the mapping of the buffer zone. For the zoning mapping, a zoning with differentiated content and rules of use was established, in line with the data and information from the previous mappings. The results obtained from the zoning proposal allow areas to be earmarked for specific purposes, and provide support for the Unit's planning and management, for drawing up the Management Plan and for decision-making.

Key words: National Forest, Environmental Zoning, Conservation Unit.

Chapter 1
1 - INTRODUCTION

With growing environmental concerns and commitments to increase the percentage of territory occupied by protected areas, the Brazilian state has been creating new conservation units, the vast majority of which are for sustainable use. However, conservation units are not only made up of scenic beauty and tranquillity, as many of these territories protected by the state are at the center of disputes over the interests of different groups (BARBOSA, 2013). This is because the creation of new conservation units, or even the expansion of existing ones, has often disregarded previous human occupation, from which socio-environmental conflicts emerge.

This problem was identified as early as the creation of the world's first national park, Yellowstone National Park, created in 1872, with an area of 800,000 hectares in Wyoming. Yellowstone had a high social impact, as it was initially designed to protect nature and the Indians, but the prevailing view after its creation was that these traditional peoples belonged to lower classes and should be expelled (DIAS, 2010).

The bibliography indicates that the Indians left willingly, "but subsequent data suggests that there were violent conflicts between the park authorities and the Shoshones: around three hundred Indians were killed in clashes in 1877, and nine years later the administration of the park was transferred to the US army" (COLCHESTER, 2000, p. 232).

According to Colchester (2000) around 70% of the world's protected areas are inhabited. In Latin America, this figure is even higher: 86%. Thus, it can be said that the social and cultural conflicts that occurred with local populations when protected areas were set up on continents such as Africa, Asia and Latin America in the first decades of the 20th century have been a harsh reality (DIEGUES, 2004).

This has been a chronic problem in the creation of nature conservation units in Brazil, too, since even the Itatiaia National Park (PARNA Itatiaia), Brazil's first protected area, created in 1937 by Getulio Vargas, still suffers from land conflicts related to human occupation of federally-owned areas.

Thus, the process of creating new conservation units has exposed the weaknesses of the National System of Conservation Units (SNUC) to the extent that even the oldest ones still suffer from many environmental conflicts, mainly related to territorial disputes and the expansion of urban areas. On the other hand, the implementation of the SNUC has broadened the ways in which society participates in public policies and in the management of conservation units, mainly through the management councils in the conservation units themselves (IBASE, 2006).

According to Barbosa (2013) "a peculiar process of social participation took place in 2008, when the category of a federal conservation unit of integral protection, the Pontoes Capixabas National Park - PNPC, was changed so that it would be possible to keep thousands of inhabitants inside". This change in category - from full protection to sustainable use -, until then unprecedented in Brazilian environmental legislation, opens the door to new interpretations and possibilities for other new processes aimed at "re-categorizing" conservation units in order to resolve socio-environmental conflicts.

In this way, it can be seen that simply establishing conservation units with the aim of halting human activity to allow the recovery of natural attributes may not initially fulfill its role effectively, especially those located in metropolitan areas. It is necessary to give a voice to the population previously established in the area and to discuss a conservation unit model that is appropriate to the local reality, in order to avoid/reduce possible conflicts.

Today, the Mário Xavier National Forest (FLONA MX) is facing a similar challenge. Located in the municipality of Seropédica, in the metropolitan region of Rio de Janeiro (RMRJ), FLONA MX has seen urban sprawl towards its boundaries in recent decades. The territory of this conservation unit is not inhabited, but the population living in the surrounding area invades FLONA MX for different purposes - grazing, logging, sand removal for use in construction, religious services - which, although not so aggressive to the environment of a National Forest - which is for sustainable use - does violate legal principles since it does not yet have a Management Plan.

In addition, FLONA MX, which for a long time was focused on sustainable forest management - the production of seedlings of forest essences, moorings, cutting timber for propping up construction sites - is home to two endemic species listed in the Red Book of Endangered Brazilian Fauna. So, given the above, what are the plans for the future of this conservation unit? How can a conservation unit be maintained under such external pressure? Is the creation of a Management Plan with use zoning enough to reduce/mitigate the conflicts or is a tougher measure, such as a proposed re-categorization - in this case from sustainable use to full protection - necessary to protect the threatened species?

This research paper is structured in 7 chapters: Introduction, General and Specific Objectives, Study Area, Bibliographical Review, Materials and Methods, Results and Discussion, Conclusions and Bibliographical References.

The first chapter presents the historical process of the relationship between society and spatial organization, its trends, causes and consequences, and the first laws in Brazil to protect and preserve nature.

Chapter 2 presents the general and specific objectives that led to this research. Chapter 3 discusses the study area, FLONA MX, its potential such as the endemic species at risk of extinction and the original Atlantic forest, discussing the possible recategorization of the unit, but also its conflicts related to the historical process and the expansion of the metropolitan region, such as the construction of the Metropolitan Arc of Rio de Janeiro (AMRJ), BR-493, which has enabled the arrival of new industries in the vicinity of the conservation unit.

Chapter 4 provides a bibliographical review of concepts such as environmental fragility and environmental conflict. It also discusses territory as a geographical concept, seeking a basis for discussing protected areas and the territory of nature protection and conservation. This is followed by a presentation of the National System of Conservation Units (Law 9.985/2000), highlighting the National Forest category and drawing a parallel with the reality of FLONA MX.

Chapter 5 deals with the Materials and Methods that were used to achieve the general and specific objectives of this research, such as the use of geoprocessing programs to carry out risk mapping.

Chapter 6 presents the Results and Discussion, where the environmental zoning proposal is presented. In this regard, maps of potentialities were generated (land use and land cover map), looking at buildings, the classification of vegetation, paths and trails, as well as a map of the areas where endangered species of fauna occur - the *Physealemus Soaresi* ray and the *Notolaebias minimus* fish. In order to analyze the conflicts, a map was drawn up showing some of the peculiar characteristics of FLONA MX. To this end, the buffer zone was mapped. For the zoning mapping, a zoning with differentiated content and rules of use was established, in line with the data and information from the previous mappings.

Chapter 7 presents the conclusions and recommendations for future studies and, finally, the bibliographical references.

Chapter 2

2 . - OBJECTIVES

2.1 - General objective

To answer the questions raised, the aim of this work is to carry out a geo-environmental diagnosis in order to support the construction of a proposal for ecological-environmental zoning to form the basis of a future Management Plan for FLONA MX.

2.2 - Specific objectives

In order to achieve the general objective, it will be necessary to achieve specific objectives:

- Research the historical process of occupation of the region;
- Carry out an inventory of the potential and weaknesses of FLONA MX;
- Carry out an inventory of environmental conflicts affecting FLONA MX, relating them to the activities carried out in and around it;
- Draw up an environmental zoning map of FLONA MX.

Chapter 3

3 - MÁRIO XAVIER NATIONAL FOREST

FLONA MX is located in the municipality of Seropédica, in the Metropolitan Region of Rio de Janeiro, in an area known as the Baixada Fluminense. Seropédica is 75 kilometers from the state capital, occupying an area of 283,794 km^2 , which borders the municipalities of Itaguaí, Nova Iguaçu, Japeri, Queimados, Paracambi and Rio de Janeiro. Its population was estimated in 2017 at 84,416 thousand inhabitants by the Brazilian Institute of Geography and Statistics, making it the 31st most populous in the state (IBGE, 2017).

It is important to note that FLONA MX is bordered to the east by urban subdivisions, neighborhoods in the municipality's headquarters and rural property; to the northeast by the Presidente Dutra highway (BR-116) and the RJ-125 highway; to the north and northwest by Fazenda Águas Lindas and the Sao Miguel neighborhood; to the west by the rural property of AMB CCP2 Empreendimentos Imobiliários Ltda., the "Minha Casa, Minha Vida" housing development and rural property; and to the southeast by the old Rio Sao Paulo highway, BR-465.

The Unit is fragmented by the Presidente Dutra highway (BR 116) to the north; and by the Arco Metropolitano do Rio de Janeiro (AMRJ) highway, BR-493, to the northwest, which divides the 493-hectare area into three. In this context, the headquarters of FLONA MX is located at coordinates 23K632545.36; E7485374.26, totally within the limits of the municipality of Seropédica, in the Rio de Janeiro lowlands; in an urban area.

It should be noted that the Conservation Unit is one kilometer from the center of Seropédica and four kilometers from the Federal Rural University of Rio de Janeiro, the EMBRAPA/Agrobiology Research Center and the PESAGRO-Rio Seropédica Experimental Station, which give the municipality great importance on the national higher education and research scene.

Figure 1 - Identification sign and access to the Conservation Unit Source: Image by the author (2017).

Figure 2 - Fragmentation to the north along the Presidente Dutra highway (BR 116)

Source: FLONA MX collection

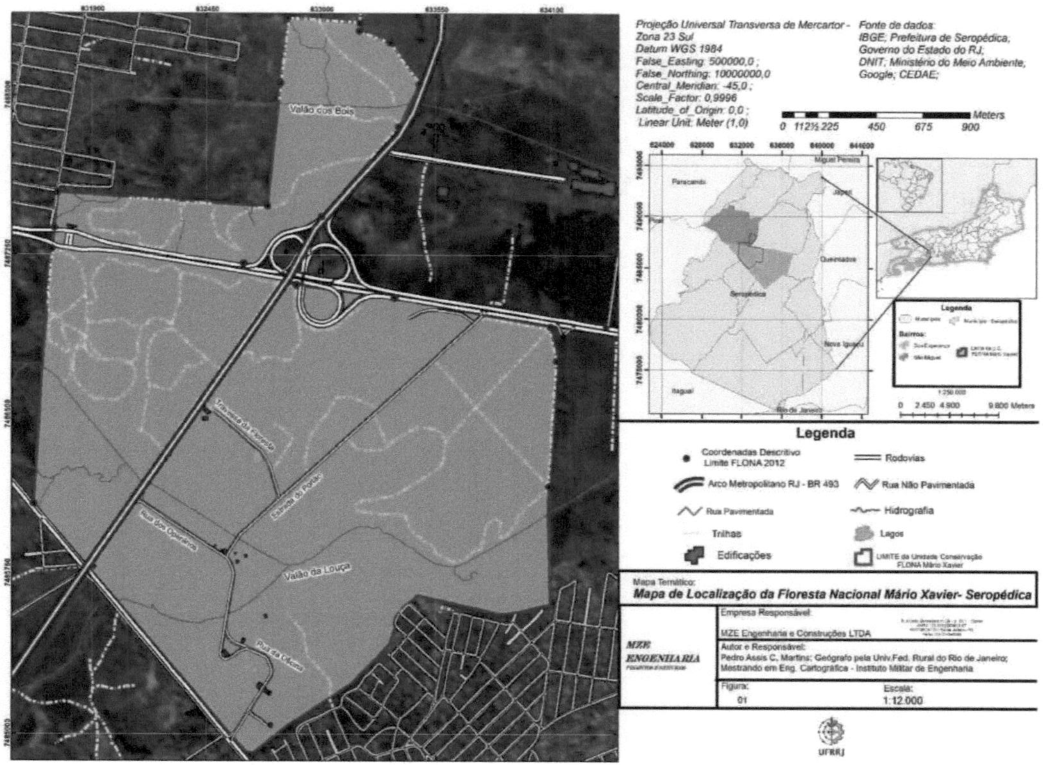

Figure 3 - Location map of FLONA MX

It should be noted that FLONA MX is a Conservation Unit (UC) established and managed by the Chico Mendes Institute, headquartered in Brasília, Federal District, where it has an organizational structure that is present in all units of the federation, represented in Rio de Janeiro by the Regional Coordination (CR8). The purpose of the Regional Coordinating Offices is to provide operational support to the management of the PAs, promote integration between the PAs and facilitate dialogue with the process coordinators in the central administration, as well as agreeing, monitoring and evaluating their performance. The specific responsibilities of the Regional Coordination Offices include promoting and supporting the actions and plans established by the Institute's management, as well as implementing its regulations. The Regional Coordination Offices were established by ICMBio Ordinance No. 07 of February 19, 2009.

It's important to remember that the initial milestone came in 2007, when the Chico Mendes Institute for Biodiversity Conservation (ICMBio) was created through Law 11.516, based on the division of IBAMA's duties. The aim of creating ICMBio was to focus on the management of federal conservation units, the protection of endangered species and speleological heritage, as well as research and knowledge management, environmental education and the promotion of ecological management. Its mission is to "protect the natural heritage and promote socio-environmental development", promoting actions to reconcile conservation with the rational use of natural resources, socio-economic development and social well-being.

IBAMA was mainly responsible for managing actions focused on the federal inspection and licensing agenda, environmental quality control, monitoring and environmental education.

3.1 - Potential

FLONA MX has a wide range of potential, including its infrastructure, fauna, flora and advisory council.

- Infrastructure:

With regard to infrastructure, FLONA MX has a wealth of movable and immovable assets of great importance in supporting the development of the Unit, with facilities and buildings such as the Administration headquarters, the living center, the house administrator's house, a technician's house, a forest nursery with a seed processing shed, a pricking-out shed, masonry beds, a carpentry workshop with machinery, a shed for storing materials, a saw mill, a large garage, a mechanical workshop, a blacksmith's shop, a heavy machinery room, a cold room and a large seed laboratory, as well as vehicles, machinery and equipment.

Figure 4 - Administrative headquarters - cold room - living center - forest nursery
Source: FLONA MX collection.

Figure 5 - Carpentry workshop - garage shed - mechanical workshop Source: FLONA MX collection.

- **Fauna:**

With regard to the fauna, it can be said that the stretches of exuberant native forest are fragmented and restricted, undergoing an uninterrupted process of ecological degradation and gradually being transformed into secondary forests with varying degrees of succession. In these extremely altered communities, most of the tree species with commercial value or greater ecological significance have disappeared (COIMBRA FILHO, 1977). Currently, the secondary forest and floodplain ecosystems stand out for their ecological importance in FLONA MX and, based on these, we have tried to highlight the most representative species, both from the point of view of rarity and economic importance, hunting, abundance, beauty, etc.

Previous studies - Santos (1999), Endemic Species Monitoring Report (SKILL, 2015), Forest Cover Inventory and Assessment (AMORIM, 2007) - aimed at surveying flora and fauna, as well as the ecological niches that are home to plant species of relevant botanical interest, have indicated that the small pools and lakes that occur in the flooded fields scattered throughout the FLONA MX area function as ecological refuges, providing shelter and food for a very diverse aquatic and semi-aquatic fauna. Two endemic species occur in these areas: the annual fish of the genus *Notholebias minimus* and the amphibian of the species *Physalaemus soaresi.* Both species are listed in the Red Book of Endangered Brazilian Fauna (MACHADO *et al*, 2008).

Annual fish are known as such because they live in temporary collections of water and produce desiccation-resistant eggs, which are buried during the mating ritual. During the dry season, the ponds and lagoons partially dry out and the fish die, the eggs remaining buried until the next rainy season, when they hatch, starting a new life cycle. Because they appeared where there was only dry land, they created the illusion that they originated in the clouds and came to the pools and ponds with the rainwater.

Figure 6 - The endemic species of Flona Mário Xavier, the fish *Notholebias mínimus* Photo: Francisco Neo CEPTA/ICMBio.

Some genera of these annual fish occur in Brazil, with *Notholebias minimus* standing out as the most widespread in terms of number of species and geographical distribution (CRUZ & PEIXOTO, 1991). The species was first discovered by General Thomas D. White and his wife in 1941 on the outskirts of Itaguaí, and was rediscovered in 1978 in FLONA MX, as reported by La Corte and NJ (1983).

The other endemic species in the FLONA MX area is the amphibian *Physalaemus soaresi*, which to date has only been found in the lagoons and wetlands within this conservation area. The protection given to the FLONA MX area since 1945 has kept alive these species representing the aquatic and semi-aquatic fauna native to the state of Rio de Janeiro, which could already be practically extinct. The survival of these species will be better ensured with proper planning of the Unit, with a management plan, including the implementation of environmental zoning that defines areas where full protection of natural resources occurs and others where sustainable use is permitted.

Figure 7 - Another species endemic to the Mário Xavier Forest, the *Physalaemus Soaresi* Photo: Ivan Borel RAN/ICMBio.

- Flora:

The information on the plant species introduced into the area and which are now present in FLONA MX, presented below, is recorded in the unit's activity control books. From the outset, FLONA MX's main activity focused on the production, marketing and distribution of seedlings and seeds. To make this possible, it was necessary to promote the first artificial formations in the area, starting small reforestations. There was a concern to form plantations of different species, with the first eucalyptus plantations (19441946). Today, the UC is home to several species of the genus, including: trabuti, robusta, saligna, botryoides, paniculata and citriodora.

With the help of the Rio de Janeiro Botanical Garden in supplying seedlings, but also through those produced on site, the planting of homogeneous stems in areas of one hectare, spaced 2 x 2 meters apart, of mostly indigenous species began (1946-1977), with emphasis on some species of the genera: Joannesia, Apuleia, Lafoensia, Adenanthera, Pterogyne, Caesalpinia, Cariniana, Falicium, Schizolobium, Lanchocarpus, Platypodium, Pinus, Erythroxylum, Dillenia, and many others.

Seed-bearing plots were also formed in 9.80 x 10.00 meter spacings of the following species: Tecoma, Tabebuia, Hymenaeae, Myroxylum, Myrocarpus, Paratecoma, Cedrela, Reputia (1952-1971). Other species appeared naturally, as did the planting of isolated species of the following genera: Enterolobium, Swietenia, Ingá, Tabebuia, Gmelinea, Puleia, Tecoma, Astronium, Cordia, Spathodea, Cássia, Crescentia, Triplaris, Ligustrum, Hibiscus, Coroupita, Erythryna, Lanchocarpus, Dalbergia, Melia, Mimusops, etc. (SANTOS, 1999).

Subsequently, following government guidelines, Technical Cooperation Terms were signed between IBAMA and Cia. Metalúrgica Barbará (now Saint Gobain), for the reforestation of 250 hectares of *Eucalyptus sp*, of which 50 hectares belong to the UC.

At the same time as the projects were carried out, FLONA MX was already exploiting and marketing forest products, such as: harvesting and processing forest seeds, producing and marketing forest seedlings, cutting and marketing logs, mills and props of *Eucalyptus sp.* in reforestation areas surveyed according to the forest inventory carried out by UFRRJ.

Figure 8 - Carving of an Atlantic forest plant species Source: FLONA MX collection.

- Advisory Board:

The Advisory Council of the Mário Xavier National Forest was made official in 2013. The first meeting was held on December 15, 2011, with FLONA staff and the advisor to Regional Coordinator 8. Mobilization took place in the first few months of the year, when four workshops were held, with the participation of some FLONA MX staff, government institutions and civil society. As a rcsult, the FLONA MX advisory council is made up of twelve institutions, until a new decree is approved/published in the Official Gazette of the Union, changing the composition of the current council, which is in Regional Coordination 8 (ICMBio, 2015).

There are currently eight representatives of government institutions and four representatives of civil society, but there have been difficulties in achieving the parity observed in Law No. 9.985/2000, Decree No. 4.340/2002 and Normative Instruction No. 11/2010, on the basis of which the council was oriented and formed. **Government Institutions**: Chico Mendes Institute for Biodiversity Conservation (ICMBIO); Superintendence of the Brazilian Institute for the Environment and Renewable Natural Resources (IBAMA); Botanical Garden of the Federal Rural University of Rio de Janeiro (UFRRJ/JB); National Research Center for Agrobiology (CNPAB/EMBRAPA); Agricultural Research Company of the State of Rio de Janeiro (PESAGRO-RIO); Municipal Government of Paracambi/RJ; Municipal Government of Seropédica/RJ and the Environment Department of the Municipal Government of Rio de Janeiro. **Civil society institutions:** Association of Rural Producers of Seropédica (APRS); Association of Rural Micro-Producers Sol da Manhä; Regional Federation of Residents' Associations of Seropédica (FRAMS); Severino Sombra University.

The council has been meeting to discuss actions regarding the activities that impact the conservation unit and its surroundings, such as the creation of the Santa Rosa Waste Treatment Center (Ciclus), the Occupation of Minha Casa, Minha Vida Condominiums, **the** Rio de Janeiro Metropolitan Arc and the Wild Animal Screening Center (CETAS).

- **Activities:**

In this rich environment, environmental education activities are carried out involving students from different educational segments of public and private institutions, both from Seropédica and neighboring municipalities. Children are shown techniques for fighting fires in forest areas and light trails that bring them into contact with nature. Young people and teenagers, on the other hand, are shown videos about Brazilian fauna and flora, the work carried out by ICMBio and trails with a higher degree of difficulty where they can see animal and plant species that are distributed in the different sections of the forest. Both groups are shown the importance of preserving and conserving biodiversity and the role of Conservation Units in complying with international agreements on sustainability and maintaining biodiversity.

It should be noted that FLONA MX's technical staff includes 6 permanent employees who carry out administrative work and environmental analysis, 24 employees on loan from another organization (Companhia Brasileira de Trens Urbanos), who carry out maintenance work, seedling production of native Atlantic Forest species and field monitoring, as well as a commissioned employee responsible for the unit's management.

3.2 - Conflicts

3.2.1 - The historical process

The historical panorama of use and occupation in the area where FLONA MX is located today is directly related to the territorial dynamics of the formation of cities, the distribution of productive land on the Santa Cruz farm and the very history of the expansion and dynamics of occupation in Rio de Janeiro.

Before occupation, the area had almost 100% forest cover, excluding some flooded lowlands, lagoons and the liquid surfaces of the larger rivers. The original vegetation consisted of Lowland Ombrophilous Dense Forest, with an emphasis on floodplain forests, typical of wetlands (MORAES *et al.*, 2013).

It is important to note that the conquest of the Baixada de Sepetiba dates back to the 16th century, in some way characterized by the first struggles against the hostile environment registered by the European settlers in search of brazilwood. The landmark of this conquest was Fazenda Santa Cruz (GOES, 1942). The economic regime, in line with the extraction of wood and firewood for local consumption, was based on livestock farming (ENGEMANN, 2002).

The legal background to the emergence of Fazenda Santa Cruz is related to a long history that begins in 1567, when Cristovao Monteiro, the mayor of Rio de Janeiro, obtained a four-league sesmaria, a survey between Sapeaquera (Itacuruçà) and Guaratiba. Half of this land was donated by Cristóvao Monteiro's widow, the then Marquesa de Ferreira, to the Jesuit priests in 1589. The other half the Jesuits obtained from Mrs. Catarina Monteiro, the daughter of the aforementioned widow, a year later, giving her land they owned in the outskirts of Santos in exchange.

Figure 9 - The Santa Cruz Farm

Source: (https://www.google.com.br/search?q=imagens+fazenda+santa+cruz+rj+antigas, accessed on 09/08/2016).

During this period, it is recorded that in 1616 the Jesuits acquired from the heirs of Manuel Veloso de Espinho five hundred bragas of testada and fifteen hundred of sertao, contiguous to their lands. It is worth noting that at the headwaters of the Guandú there was a rich sesmaria of six leagues in quadra, granted to Captain Manuel Correa and others, confronting the lands of the priests of the Society of Jesus, which, by deeds of 1654 and 1656, were acquired from the heirs of Manoel Correia, also a Jesuit.

incorporated into the farm. Strengthening the characteristics of large estates, the Jesuits gradually incorporated new lands, thus creating the grandiose territory known as Fazenda de Santa Cruz.

According to Viana (1974), the ten leagues of the farm were formed with three sesmarias, covering the lands situated from the islands of Guaraquegaba, in the Pedra of the parish of Guaratiba and Itingussu, in the municipality of Mangaratiba, to the Serra de Mata-Caes, in the municipality of Vassouras. However, started in 1729 and completed in 1731, the tombo redefined some geographical boundaries which became: the foothills of the Sacra Familia do Tinguá parish, in the current municipality of Vassouras; the Curral Falso line at the back; the boundary with the Guaratiba parish to the sea; the Marapicú lands to the east; and the Mangaratiba lands, at the Itingussú point, to the west.

Today, with its privileged location, it would correspond to the areas occupied by the entire district of Santa Cruz, in the municipality of Rio de Janeiro, and other municipalities in the state:

Itaguai, Seropédica, Barra do Pirai, Mendes, Nova Iguagu, Paracambi, Engenheiro Paulo de Frontin, Pirai, Rio Claro, Vassouras and Volta Redonda. The farm, with 2,167 km^2 or approximately 44,770 alqueires, was one of the most prosperous properties belonging to the order of priests of the Society of Jesus in the south of Brazil.

It was against this backdrop that the engineering works carried out on the site were carried out to minimize the heavy flooding, and variably related to the significant increase in farming activities around Sepetiba Bay. It's important to remember that the need for the Jesuits to use the territory effectively made it possible, through the techniques they developed, to make better use of it and consequently to develop the region. In fact, livestock farming was the main economic activity of the estates at the time of the Jesuits, complemented in the background by rice cultivation (SANTOS, 1970, p.52; *apud* MARQUESE, 1997).

Among the interventions were the construction of an earth dyke to contain the torrents coming down from the mountains, diverting them away from the River Guandu and preventing the pastures from flooding; the construction of the São Francisco canal, which diverted the waters of the Guandu into the River Itaguai, with extensive transverse drainage ditches being built to allow the water to flow towards the sea. Noteworthy here is the Jesuit Bridge dating from 1752 (Figure 9), built over the Guandu River, equipped with a system of registers and sluices that allowed irrigation of the fields during the dry season and control of the volume of water.

Figure 10 - Jesuit Bridge

Source: (https://www.google.com.br/search?q=imagens+fazenda+santa+cruz+rj+antigas, accessed on 09/08/2016)

The Jesuit priests did a lot to transform those inhospitable, unproductive and swampy lands into fertile lands that would provide them with resources for their catechetical works (GOES, 1942). The result is the complex system of drainage, irrigation and dams developed in the Guandu basin, which bears witness to the fact that the priests were responsible for the first changes to the river system in the area surrounded by Sepetiba Bay, using slave labor (CARELLI, 2012).

With the expulsion of the Jesuits in 1759, by decree of the Marquis of Pombal, the region went through a period of decay. The Santa Cruz estate was incorporated into the assets of the Portuguese Crown, then known as Fazenda Real. Neglect and successive wrong plans, abandonment, selfishness and malice brought everything to ruin. The farm's vast production gradually disappeared (VIANA, 1974).

Twenty years after the expulsion, they reorganized under the government of Viceroy Luiz de Vasconcelos e Sousa to continue with the work of the Jesuits, opening new roads and paths throughout the hinterland of the estate, beginning a new period of cultivation (SANTOS, 1999). In this sense, according to VIANA (1974), with the main objective of making progress in the production of goods and merchandise, in 1795 Colonel Manuel Martins do Couto Reis was appointed administrator of the estate. Thus, his administration began a period of prosperity, in which he implemented new improvements, including two sugar cane mills: Engenho de Itaguaí and Engenho Piauí, and new corrals in the extensive area of the farm, highlighting the development of extensive livestock farming during this period.

In 1804, for political reasons, Colonel Manuel Martins do Couto Reis was removed from the administration of the estate and the lands of the Itaguaí and Piauí mills were sold off, thus beginning a new phase of decay and stagnation. The dismissal of Colonel Manuel Martins do Couto Reis and the choice of a new administration allowed the dismemberments to continue, until in 1813 the town of Sepetiba emerged, and in 1818 the village of Itaguaí was built on the site of the old Indian village (SANTOS, 1999).

According to VIANA (1974), there is a close relationship between the dynamics of leasing and the illegal use of the lands of Fazenda Santa Cruz, since in 1786 there were already groups who owned rice plantations and other products without the legitimate permission of Superintendent Leonardo P. de Vasconcellos. On the other hand, the actions of authorized landowners were not always in line with the interests of the estate, as was the case when large areas of virgin scrubland in the region were destroyed. Thus, they were characterized as conflicts, which already existed at the end of the 18th century.

In the second half of the 19th century, with the creation of different laws that gradually led to the abolition of slavery, the lack of labor for rural services led to the abandonment of vast areas. The lack of maintenance throughout the existing water network caused drainage channels to become clogged, flooding large tracts of land, making it difficult to use them, and malaria made the Baixada de Sepetiba region depopulated and abandoned (GOES, 1942).

In 1889, with the proclamation of the Republic, Fazenda Santa Cruz was renamed Fazenda

Nacional de Santa Cruz, under the subordination of the Directorate of Public Revenue (SANTOS, 1999). However, during the first years of the Republic, the region remained unpopulated and practically abandoned, just as its first explorers had found it (VIANA, 1974).

From 1891 onwards, people began to think about cleaning up the Baixada de Sepetiba, but it was only after 1918 that governments began to take the region's problems seriously (SANTOS, 1999). All attempts, however, failed, whether due to a technical misunderstanding of the problem, the lack of detailed studies, the lack of unity of direction, administrative discontinuity or insufficient funds (GOES, 1942).

Due to the recurring failures, President Getúlio Vargas put in place a general plan with well-defined guidelines to promote the desiccation of flooded lands. The cleaning and rectification of the drainage channels of the Baixada de Sepetiba and the consequent extinction of countless swamps was decisive in reducing the outbreaks of malaria. Once the unhealthy conditions had been remedied, people were once again able to settle, opening up new prospects for the region. The process of improvement through accessibility via roads allowed for the emergence of new buildings, new fields of culture and the implementation of other colonial centers (GOES, 1942).

According to Magnanini (*apud* SANTOS, 1999), at the beginning of the 19th century the new roads made it possible for agricultural and urban allotments to spring up, completely altering the natural landscape. The extensive pastures, the source of occasional fires, the crops and the allotments left almost nothing of the primitive vegetation cover.

In the 1940s, President Getulio Vargas, determined to boost economic development in the region, created several rural settlements on land that belonged to the Fazenda Nacional de Santa Cruz, including the Núcleo Colonial de Santa Cruz, with the aim of developing agriculture and livestock farming in the region. These activities led to the establishment of the National Center for Agronomic Teaching and Research, which was subordinated to the National School of Agronomy and the National Veterinary School, the precursors of the current Federal Rural University of Rio de Janeiro.

A new phase of activity began, favored by the proximity of the BR-465, the old RJ-SP highway, the main link between the two capitals until the 1950s, when the BR-116 was inaugurated, better known as the Presidente Dutra Highway, which is still today the main link between the country's two largest capitals.

This period saw the creation of the Horto Florestal, strategically located near the National School of Agronomy, whose main objectives were to experiment, demonstrate and disseminate silvicultural practices, but also to produce seedlings of native and exotic species.

Figure 11 - Sowing seeds in the background Santa Cruz Forest Garden Source: FLONA MX collection.

The Horto Florestal was subordinated to the Forestry Section of the Servigo Florestal, an organization linked to the Ministry of Agriculture, by Decree No. 9.015 of 16/03/1942. Getulio Vargas had recently created the Forestry Service with the aim of protecting the country's forests, guarding and conserving them, in accordance with the Forestry Code of 1934, in force at the time, as well as promoting forestry and organizing national parks, forest reserves and typical forests (BRASIL, *apud*, SANTOS, 1999:40).

Figure 22 - Territory of the Horto Florestal de Santa Cruz - 1940s

Source: FLONA MX collection.

In June 1945, agronomist Mário de Figueiredo Xavier, an employee of the Ministry of Agriculture, took over as administrator of the Horto Florestal. At this time, many people were already living in the 27 houses located in the garden, all of them employees with their families. Including the Horto administrator with his family.

Figure 33 - Dr. Mário Figueiredo Xavier - Arbor Day Source: FLONA MX collection.

On September 15, 1945, President Getulio Vargas unveiled the inaugural lake of the Horto Florestal de Santa Cruz, with an area of approximately 1,250 hectares.

Figure 44 - President Getulio Vargas at the inauguration of the Santa Cruz Forest Garden

Source: FLONAMX collection.

Initially, Mr. Mário Xavier was concerned with cleaning up the officially delimited territory, promoting drainage and the flow of water that had accumulated on the land. This action encouraged the planting of fast-growing forest species such as *Eucalyptus*, which were widely encouraged at the time, endorsing the experimental work carried out by the Companhia Paulista de Estrada de Ferro (SANTOS, 1999).

Figure 55 - Santa Cruz Forest Garden Road - Late 1940s
Source: FLONAMX collection.

The administrator, Mr. Mário Xavier, remained in charge of the Horto until 1951, when he was transferred to the Agricultural Development Section of the Forestry Service, in the district of Campo Grande - DF, where he remained until 1954. It should be noted that in the same year he returned to the Santa Cruz Forestry Garden as head, which was renamed the Santa Cruz Experimental Forestry Station.

In addition, at the end of the 1950s, around 750 hectares of the original Horto Florestal area were given over to the establishment of the Núcleo Colonial de Santa Alice. As a result, the Santa Cruz Experimental Forest Station was reduced to approximately 500 ha.

The Santa Cruz Experimentation Forest Station was renamed the Mário Xavier Agronomist Experimentation Forest Station in honor of its administrator, Dr. Mário Figueiredo Xavier, on September 27, 1970.

Figure 66 - Plaque in Honor of Dr. Mário Xavier Source: FLONA MX collection.

Federal Decree No. 93.693 of 08/10/86 transformed the Engenheiro Agrónomo Mário Xavier

Experimentation Forest Station into the Mário Xavier National Forest, with economic and social purposes, covering an area of approximately 493 ha, under the administration of the Brazilian Forestry Development Institute - IBDF, which was later transformed into the Institute for the Environment and Renewable Natural Resources - IB AMA.

It is important to note that the practice of planting fast-growing forest essences (Eucalyptus) was not only implemented by agronomist Mário Xavier at FLONA MX. In 1997, Companhia Metalúrgica Barbará, which preceded Saint-Gobain Canalización, signed a 23-year loan agreement with IBAMA, resulting in the implementation of a reforestation project, with eucalyptus species, on 250 hectares of FLONA MX - more than 50% of the federal public area - in which it was agreed, among other obligations, that IBAMA would be entitled to 25% of each cut of the forest massif, which would be sold to the company itself.

However, in 2007, following the creation of the Chico Mendes Institute for Biodiversity Conservation, a federal agency that succeeded IBAMA in the role of managing federal nature conservation units (art. 3 of Law 11.516/2007), the deals signed between the former authority and the steel company were rescinded, following IBAMA's own guidelines, since the loan, the exchange and the technical cooperation agreement were signed without prior tendering and in disagreement with current federal legislation.

However, undoing the agreements did not solve the problem, as the eucalyptus trees are still in the MX FLONA. ICMBIO and Sant-Gobain Canalizaçao negotiated a Conduct Adjustment Agreement, which did not reach a successful conclusion due to disproportionate proposals made by the company, according to technical information prepared at the time by FLONA MX.

Today, twenty years after the project was implemented by Sant-Gobain Canalizaçao, its lawyers, in a conciliatory hearing, expressed the company's interest in giving up its claim to the investments made in the FLONA MX area, leaving the eucalyptus stalks in ICMBio's ownership. ICMBio, however, contested the company's initiative, claiming that it was unfeasible to just return the eucalyptus, as it believes that there should be an advantage for FLONA MX - ICMBio, and in this sense it is demanding that a forest inventory be carried out, the eucalyptus removed and native species of the Atlantic Forest reintroduced, with a 5-year maintenance period.

It is interesting to note that many of the small environmental conflicts still found in FLONA MX today are remnants of practices permitted at other times in history, when the area was not yet a conservation unit, whether it be cattle breeding, removal of vegetation and cutting down of trees, night hunting or even the sweeping up of fallen leaves in forested areas for the practice of religious services.

There are also eucalyptus trees that were introduced when FLONA MX was still the Horto Florestal, and those introduced in 2007 under the management of IBAMA, under the name FLONA

MX, and which today weaken the maintenance of the area because they are characterized as an invasive species. Eucalyptus is a species native to Oceania and Southeast Asia, and therefore does not belong in the Brazilian Atlantic Forest. There is a lot of controversy surrounding the Eucalyptus. Despite their high economic value, they are able to grow quickly and absorb large amounts of water, which is a competitive advantage over other native plant species, leading to a loss of biodiversity.

The presence of exotic species in a sustainable use conservation unit such as a National Forest is not characterized as an environmental conflict (Art. 31 § 1 of Law 9.985 of July 18, 2000), however, there must be forest cover of predominantly native species (Art. 17 of Law 9.985 of July 18, 2000). 2000), which is not the case. Thus, the conflict exists not because of the presence of eucalyptus, but because of the extent that it occupies, mainly due to the contract established between IBAMA and Saint-Gobain Canalizaçao. Thus, everyday practices began to generate disorder and were considered conflicting just when the conservation unit was created with the intention of establishing order, which demonstrates the complexity of the situation (MORIN, 2005).

Today, FLONA MX is widely used by companies based in the municipality of Seropédica, which request an area where they want to implement ecological reforestation projects from environmental compensation, which may or may not be accepted by the UC manager. In this sense, by having a public area that can be planted, companies save a lot with the existence of this PA, and its permanence is fundamental to the private sector directly and to the state indirectly, since if there is no public area for forest replanting projects, they need to be implemented in private areas, making them very costly and sometimes unfeasible.

On the other hand, for the Seropédica City Council, the MX FLONA has proved to be an obstacle to urban expansion, given that urban land in Seropédica has been highly valued since the start of the works for the implementation of the Metropolitan Arch (BR 493), and it has already requested the municipalization of this federal public area on more than one occasion. One was through the opening of a process with ICMBio with the justification that there would be investments by the municipal government to make the area more attractive and better used by the population. Another more recent one, in 2012, through the Seropédica Municipal Strategic Planning, where FLONA MX appears with urban interventions inside it, such as the construction of a lake with kiosks to sell products from the region, for example.

3.2.2 - The implementation of the Rio de Janeiro Metropolitan Arc

The construction of the Rio de Janeiro Metropolitan Arc (AMRJ) is justified as a project of special interest for the state of Rio de Janeiro, for the Southeast region, but also for the country as a whole. Its route, which passes through the Baixada Fluminense, on the outskirts of the city of Rio de

Janeiro, relieves access to the RMRJ, increasing accessibility in this area and significantly reducing operating costs and travel time for vehicles passing through the region.

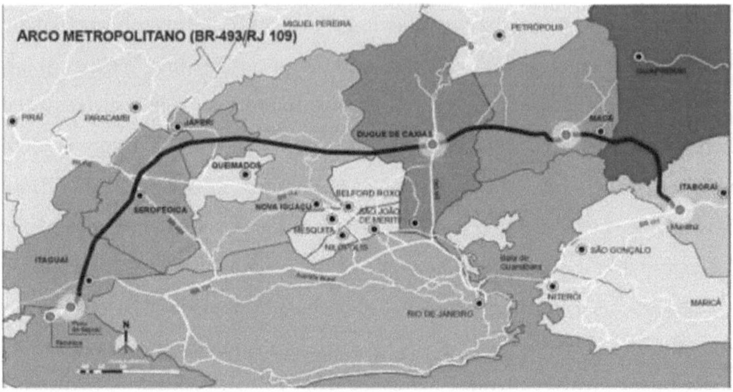

Figure 77 - Rio de Janeiro's Metropolitan Arc Trail

It is important to note that for the Federal and State Governments, this project is of fundamental importance for long-distance national freight transportation, as it interconnects the BR-101 North (Itaboraí) and South (Itaguaí) highways, as well as the BR-040 (Rio-Belo Horizonte) highway without passing through the state capital, especially Avenida Brasil and the Rio - Niterói Bridge. It also aims to connect the industrial areas of the Duque de Caxias Refinery and its support centers, Comperj (Itaboraí) and Congás (Maricá), among other industrial districts in the RMRJ to the Port of Itaguaí, which is currently severely hampered by the precarious conditions of its road access, one of the main factors inhibiting its development.

The project to build the AMRJ dates back to the 1970s, when the former states of Rio de Janeiro and Guanabara merged - the first and only such merger in Brazilian history - but it was not carried out well due to the failure of the geopolitical project of Brasil Potencia (EVANGELISTA, 1998). They united two very different states economically, which created integration problems and aggravated the crisis that had already hit Rio's economy in the 1970s and 1980s, due to the loss of the federal capital (SANTOS, 2003).

Thus, the AMRJ project came off the drawing board more than forty years after its conception to fulfill its objective of boosting Rio's economy. It was made possible by the political alignment between the federal and state administrations, which hadn't happened for decades, which made its implementation financially viable, and negotiations began in 2003. The AMRJ was included in the federal government's Growth Acceleration Program (PAC), ensuring the necessary financial resources for its construction, delegated by agreement to the state of Rio de Janeiro.

In order to move forward with the project in 2008, the Secretary of State for Works

(SEOBRAS) and the Department of Roads Foundation (DER) began work on the implementation of the AMRJ, BR-493, with the contracting of a consortium to begin civil construction, a fact that has already caused some disruption to the management of FLONA MX.

The implementation of the works on the stretch of the conservation unit was delegated through the presentation of Environmental Impact Studies/Environmental Impact Report (EIA/RIMA), as the project is non-conservationist in nature and has the potential to generate negative impacts on the unit. The EIA/RIMA considered four alternative locations for the project (1A, 1B, 2 and 3), assessing the socio-economic, environmental and cost aspects of the work (vol. I, chap. IV, pg. 2/12).

The EIA considered alternative 1A to be the most viable, which involves the highway crossing the old RJ-109, which crosses the MX FLONA - currently not used as a traffic route, but only as an internal access route to the FLONA's facilities. It also evaluated the highway overpass (viaduct along the entire MX FLONA crossing) on this same road (alternative 1B), which was not considered to be the most viable alternative.

It's worth noting that both alternatives, using the old RJ-109 road, affect IBAMA's Wild Animal Screening Center (CETAS), located inside FLONA MX. This CETAS is considered a regional reference in the reception, treatment and reintroduction of wild animals seized in the state of Rio de Janeiro, and has received significant financial investment from Parliamentary Amendments in the context of the CETAS Brazil Project. As such, any change in its structure or even the need to relocate it will require inter-institutional coordination, as well as an appropriate analysis of technical feasibility.

Location alternative 2 has a similar or lower potential environmental impact for FLONA MX, despite requiring the relocation of around 120 families. It could represent socio-economic gains for the impacted population, has an estimated cost that is intermediate between the proposals, and does not require the relocation of the CETAS.

In terms of location, alternative 3 does not intersect FLONA MX, but it does involve the relocation of the Viúva Graça toll plaza on the Presidente Dutra highway, as well as 250 homes. The relocation of these families, considered low-income, mentioned in alternatives 2 and 3, could represent a positive socio-economic aspect in that they could be resettled in neighborhoods with better urban infrastructure, transport and even job offers

Although it was not included in the EIA/RIMA, in contact with the head of FLONA MX, it was reported at the time that there was a fifth proposed alternative location, consisting of the same route considered in alternatives 1A and 1B (site of the old RJ-109), but the crossing over the FLONA would take place as follows: (a) a 400 m stretch starting from the viaduct over the Presidente Dutra Highway, (Japeri Viaduct) suspended over the FLONA; (b) a 400 m stretch of grating laid on the ground inside the FLONA and (C) a further 400 m of suspended grating crossing the rest of the

FLONA and consisting of the crossing over the BR-465. Although this alternative apparently proved promising and had the potential to minimize impacts on the Unit, it continued to affect the CETAS, with the implications already mentioned. It is worth noting, however, that this proposal was not formally presented in the EIA/RIMA, making it necessary for this Institute, which is responsible for the management of FLONA MX, to analyze it. References to this fifth proposed alternative location were apparently made in the penultimate paragraph of SEOBRAS/SSURM Official Letter No. 064/2007 of 08/09/2007, which forwarded the EIA/RIMA, but without further details.

It is important to emphasize that, according to the concessionaire's analysis, which evaluated the implementation of the work taking into account the lowest cost, with three variants as an alternative, the "segment C" route was chosen in which the road crossed the MX FLONA.

In view of the above, the environmental impacts were initially caused by the movement of earth and equipment, especially in the extraction of gravel in the borrow area of the E-15 deposit, located near the intersection of the AMRJ and the Presidente Dutra highway, BR-116. This area is approximately four hectares of undulating terrain, commonly known as "half orange". A

The topographic surface of the area was already quite altered in terms of soil conditions and vegetation cover.

Despite the adverse environmental conditions, there was a wealth of water resources in the form of four springs covering the entire glebe and becoming, from a legal point of view, a "Permanent Preservation Area" - APP. However, with the mineral extraction works, the glebe suffered major environmental alterations due to the use of heavy equipment to remove large volumes of gravel to fill in the AMRJ's right of way, along the stretch that cuts through the MX FLONA. To carry out the works, it was necessary to remove the vegetation cover, remove the earthy material, culminating in the lowering of the relief to the gravel structure, transforming practically the entire site into a rocky topographic surface, which caused major interference in the flow of water, such as new drainage corridors and the intensity of these flows.

Thus, the area that was classified as "Altered Area" has been transformed into "Extremely Degraded Area", which will require a much longer time to recover. Revegetation is necessary, with the replacement of a properly protected substrate and the implantation of native species of Atlantic forest, which will allow greater infiltration of water into the soil, providing a more active recharge, as well as favoring the return of animal species that used the area as their habitat.

Figure 88 - Area used as a source of mineral extract used in the AMRJ preparation

Source: FLONA MX collection; 2012.

In addition to the conflicts caused by the construction of the AMRJ in FLONA MX, there are other impacts caused by the existence of this highway, such as:

- internal pathways obstructed by the drying out of the unit;
- lack of access to firebreaks for maintenance and fire prevention;
- a strong impact due to the proximity of the road to IBAMA's CETAS;
- environmental pressure on endangered species;
- car accidents in AMRJ that fall within FLONA MX.

As this is a conservation area, the road must be maintained:

- specific signage project with indicative, educational and warning signs;
- risk management program and emergency action plan for setting up emergency response units for accidents involving dangerous chemical products; - program for monitoring and maintaining road drainage systems;
- sound barrier installation project.

The projects and programs listed were presented by the developer in the form of a report describing the specifications for each of the items which, however, have not yet been installed. For example, the acoustic barrier was identified as necessary after an acoustic modeling and computer simulation study was carried out, as required by a Technical Note from ICMBio/RAN. It should be noted that the study followed the relevant technical standards and legislation, such as: Associagao Brasileira de Normas Técnicas. NBR10151: 2000-Acoustics - Noise assessment in inhabited areas, aiming at community comfort - Procedure. ABNT, Rio de Janeiro, 2000; ABNT NBR 14313:1999 - Noise barriers for traffic routes - Construction characteristics.

In view of the expected heavy daily traffic of light and heavy vehicles in AMRJ, the purpose of the acoustic barrier is to reduce the noise impact on FLONA MX, limiting the noise allowed during the day to 40 dB (A) and 35 dB (A) at night. The noise barrier should be installed as close as possible to the roadway, so that it works as efficiently as the noise simulation.

Another problem is the insufficient presence of indicative, educational and warning signs, coupled with the lack of speed enforcement, as well as the lack of metal fencing along the entire roadbed. The combination of these factors means that high-speed vehicles are involved in accidents and are dragged into FLONA MX. These accidents highlight the urgent need to implement the other two programs mentioned, the risk management and emergency action plan, to set up emergency response units for accidents involving dangerous chemical products, and the program to monitor and maintain the road drainage systems.

Figure 99 - There is no fixed or mobile speed control or metal fencing on the road Source: FLONA MX.

Figure 20 - Truck at the scene of the accident, inside FLONA MX (Source: Seropédicaonline.com - accessed on 23/10/2016).

On this subject, it is worth transcribing an article published in the newspaper O Globo, in Ancelmo Gois' column, on November 6, 2014, in which he reports: Look at the photo. "The car rolled downhill on BR-493, the Metropolitan Arch, at the height of the municipality of Seropédica. It is the fourth vehicle to crash there in less than a month because the road, which connects the Port of Itaguaí to Comperj, has no protection barriers along this stretch, contrary to what was agreed with INEA. Accidents have occurred in the area of the Mário Xavier National Forest, and the noise of the engines frightens the animals at Ibama's Triage Center. It was even planned that the barriers would have an acoustic system so as not to disturb them. May God protect the green, free the animals from this scare and never abandon us."

In this sense, with a view to minimizing direct and indirect impacts on the conservation area, we are witnessing the advance of economic planning, in which the state of Rio de Janeiro is integrating the expansion of the metropolis on multiple scales through new economic, industrial and logistical centralities located on its borders (OLIVIERA 2015). As such, there has been an increase in environmental licenses for new developments that are arriving in the area surrounding FLONA MX. In addition, according to the same author, in the last 10 years there has been an increase in investments aimed at integrating the metropolitan region, with the implementation of the AMRJ and the strong attraction of the port dynamics of Itaguaí.

It's important to remember that in this decade there have been important investments in the municipality of Seropédica, attracted by the implementation of the AMRJ, and the city council has announced in the local media the installation of 8 logistics condominiums. The highway network, with major expressways, reverberates the receipt of productive inputs on its banks and is aimed at the economic development of the region. Consequently, at this time, several mixed companies have been setting up, encouraged by the reduction in costs and, concomitantly, by the creation of the industrial district, which already has several enterprises - Procter & Gamble, Eletrobolt (recent expansion), Panco, Brasilit, Merfetec, BRF Brasil, which also includes the creation of a UFRRJ Technology Park, which will be installed in Morro das Pindobas, on the banks of the BR-116, Presidente Dutra highway.

However, Oliveira (2015) points out that Seropédica, through these investments in industrialization, albeit preliminary logistical reordering of the territory, is transformed from a city that just a few years ago had "rural" features, and progressively breaks away when it receives the BR-493 highway, AMRJ. As potentially polluting companies set up in the region, it is liable to receive a greater pollution load, of air, water and soil.

In view of the above, it is worth noting that there are countless challenges to be overcome in order to improve the management of FLONA MX as a result of the fact that the AMRJ works are in their final phase, in accordance with the Installation License (LI n° IN031395). The license renewal is valid until August 12, 2017, issued by decision of the Board of Directors - CONDIR (INEA) at its 298th Ordinary Environmental Licensing Meeting.

Chapter 4

4 - LITERATURE REVIEW

4.1 - Environmental fragility

Before the intense technological development of our society, it could be said that, with the exception of a few areas of the planet, the Earth was in equilibrium, a dynamic equilibrium. With the passage of time, the scale reached by man's capacity for derivation in nature became such that, as early as the 1990s, Monteiro (1992, p. 10) considered it to be equivalent to "the greatness of the work of the forces of nature themselves". Thus, "the fragility of natural environments in the face of human intervention will be greater or lesser depending on their genetic characteristics" (ROSS, 2006, p.50).

Thus, from the second half of the 20th century onwards, the interdependent relationships between physical-environmental aspects and anthropogenic activities began to be investigated through multi-thematic assessments (SANTOS, 2015). This can be seen in the works of Tricart (1977) on the transformations that society promotes in landscapes and the constitution of ecodynamic units; Sotchava (1978) on the concept of Geosystems; Ab'Saber (1994) who presents the concept of Total Space; Ross (2006) when he develops the Ecogeography of Brazil; Christofoletti (1999) addressing the modeling of environmental systems, and many others.

Studies related to environmental fragility are extremely important for environmental planning. The identification of natural environments and their potential and emerging fragilities provides a better definition of the guidelines and actions to be implemented in the physical-territorial space, serving as a basis for zoning and providing subsidies for land management.

A tool widely used by various public bodies to draw up environmental territorial planning is the creation of environmental fragility maps. This tool makes it possible to assess the potential of the environment in an integrated way, matching its natural characteristics with its constraints.

The empirical fragility methodology proposed by Ross (1994) is based on the principle that nature has intrinsic functionality between its physical and biotic components. The operational procedures for its construction first require basic studies of relief, soil, geology, climate, land use and vegetation cover, etc. Subsequently, this information is analyzed in an integrated manner, generating a synthesis product that expresses the different degrees of fragility that the environment has as a function of its genetic characteristics.

4.2 - Environmental conflict

Environmental conflict is understood to be one in which there is a clash of interests represented around the use and/or management of the environment (CARVALHO and SCOTTO, 1995, *apud*, LOUREIRO et al., 2008). Environmental conflicts can be categorized as explicit or implicit.

Implicit conflicts are those in which communities are affected by a process of degradation of which they are unaware. In some cases, even if they are aware of it, communities do not associate environmental degradation with specific social practices and agents. Environmental conflict is made explicit when communities establish an immediate logical association between environmental degradation and the actions of specific social agents under given historical conditions (ACSELRAD, 1992).

Following a procedural classification made by the National Forum for Colombia Foundation (2001), there are three stages in a conflict: the emergence of an incompatibility (latent conflict); becoming aware of this opposition (incipient conflict), and taking a position to confront and overcome it (manifest conflict).

It also highlighted the need to realize that within this context, environmental conflict can involve social groups with different ways of appropriating, using and signifying the territory. There is a conflict when at least one of the groups has the continuity of their ways of appropriating the environment threatened by undesirable impacts, transmitted through the soil, water, air or living systems, resulting from the practices of other groups.

Conflicts can arise from disputes over the appropriation of the same resource base or of different bases, but interconnected by ecosystem interactions mediated by the atmosphere, the soil, the water, etc. The type of conflict we have identified here has as its arena the same territorial unit shared by a set of activities whose symbiotic agreement is broken as a result of the denunciation of the undesirable effects that the activity of one of the agents produces on the material conditions for the exercise of the practices of other agents.

Environmental problems and social inequalities place populations in situations of risk and vulnerability. Thinking about populations affected by environmental conflicts means considering the legitimacy, ownership and use of territory, as well as the ways in which these spaces are occupied. The conception of territories involves the incorporation of practices that affirm place, through multiple constitutive practices.

It's worth pointing out that environmental conflict occurs through expropriation or environmental impacts that prevent communities excluded from the development project from maintaining their socio-cultural and socio-environmental forms of reproduction. These communities have different ways of looking at the world that oppose the big projects. They are usually quilombolas, indigenous groups, small farmers, etc.

Recognizing that society is made up of conflicts does not mean, from a democratic and dialogical perspective, that it is impossible to negotiate and seek consensus that results in the resolution of an identified problem, but rather to understand that agreements take place on a conflictual basis in which we move and that the ability to seek dialogue, rapprochement and respect defines the maturity of the social and political participation that exists in a society.

Therefore, the management of FLONA MX, in agreement with the Regional Coordination (CR8), has sought to develop environmental protection actions in order to monitor practical conflicts and their negative effects on the Conservation Unit. Urban expansion and major construction projects are the biggest pressures on the conservation area, not only because of the economic growth around the new Port of Itaguai, but also because of the presence of allotments and neighborhoods on the southeast, northwest and west borders of FLONA MX, which imply conflicts and damage when they occur within the boundaries of the conservation area: sewage disposal, garbage disposal, shitting, fishing, logging, livestock presence... Consequently, there is a clash of interests represented by different social actors in relation to the protected area.

4.3 - Protected areas: territory for the protection and conservation of nature

Protected areas are parts of the territory under special care and attention, in natural and semi-natural spaces, geographically defined, regulated and managed with the objectives of conservation and sustainable use of biodiversity. It is estimated that natural areas in the world, especially protected areas, total around 32.6 million square kilometers, which is equivalent to around 6.4% of the Earth's surface. Many authors consider these areas to be the main tool for protecting biological diversity, and their creation and management are important components of any environmental policy (DUDLEY, 2005).

However, every year the world's population increases and, consequently, the demand for natural resources is causing serious environmental problems. This, coupled with the intensification of production activities around protected areas - industrial, agricultural and mining - means that the degradation of natural resources is reaching a critical stage all over the planet. Investment in economic, technical and scientific resources in the planning and management of protected areas is therefore essential in the quest for sustainable use of natural resources.

In view of the above, FLONA MX is therefore part of a region whose socio-economic dynamics tend towards population growth, increased pressure on natural resources (water resources and biodiversity), and an increase in mining activities as an economic alternative. The strong tension between the demands for protection of the natural environment and those for economic development in the region therefore seems evident. This calls for a new concept of management capable of aligning biodiversity conservation strategies with the social demands arising from the socio-economic dynamics of the area.

Society has recently become more concerned about the use of natural resources, which is growing in the current prevailing model of the relationship between man and nature. In this sense, geography as a science allows us to glimpse the concept of territory with greater balance and awareness by treating the geographical space as an infinity of possibilities. It allows the subjects participating in this process to have various relationships, whether individual or collective, in shaping these territories.

Modern geography, specifically scientific geography, focuses on and discusses the organization of man in physical space - the physical elements of nature, seeking to understand the relationship between man and the natural environment, projecting the relations of power in space (SOUZA, 2000). It should be noted that the way in which society appropriates and uses space is guided by principles, often based on its dynamics, since the very modification of space is the concreteness of the process of authority over it. According to Rocha (2014), territory is delimited, constructed and deconstructed by power relations, and its exercise involves a growing scale of various actors who territorialize their actions for different periods.

It is important to note that Brazil's territorial management policy gained prominence with the creation of the National System of Conservation Units (SNUC) in July 2000, outlining the concept of a territory legally instituted by the government, with its own characteristics and institutionally defined regime. Unconditionally, since the creation of the SNUC, environmental and territorial conflicts have been the subject of various studies centered on the interfaces between human populations and protected natural spaces (DIEGUES, 1996).

We note here that among the criteria and rules established for the creation, implementation and management of conservation units in Brazil in an integrated manner, the objectives are to contribute to the maintenance and restoration of the biological diversity of natural ecosystems, promote sustainable development and protect natural and little altered landscapes of outstanding scenic beauty (BRASIL, 2000).

We therefore point out the importance of this approach to the delimitation of territory for nature conservation, its motivations and the means used with the dynamics of society in the sense of recording what has been lived, both the trajectory of FLONA MX and the memory of the surrounding population. In this context, the boundaries of this territory are being drawn by frontiers which show man's relationship with a portion of space, with a defined location and limits, where they have important ecological characteristics.

This means that it is through the pragmatic changes experienced, lived and especially those directly suffered by the people living in the allotments around the conservation unit, where the territory is under strong anthropic pressure (areas bordering the allotments), that we can gain a better understanding of society and its relationship with nature. Bearing in mind that the emergence of new territories and territorialities can inevitably bring conflicts between the parties involved, it is necessary to remember that conflicts have existed since the dawn of society. They enable the process of human evolution and are even necessary for the development of any social, political and organizational system.

Conflicts can arise from the very beginning of the changes resulting from the loss of an area built on affective, symbolic and identity relationships. In this way, we agree that "men are central to the formation of each territory: crystallizing relationships of influence, affective, symbolic, conflicts, identity, etc" (SAQUET, 2009, p. 81). As such, these aspects are fundamental in bringing the concept of territory closer to understanding the purpose and role of creating protected areas.

Historically, the justifications and motivations surrounding the creation of protected areas have varied according to different visions of nature and the importance of protecting it. Thus, these aspects reflect the cultural, social, political and scientific expressions in force in each situation and time (FRANCO; SCHITTINI, 2010). Protected areas are territorial spaces where human use, at different levels and for different purposes, is legally limited or prohibited. In view of this, the first human societies already delimited areas with a view to preserving their natural attributes, which could be directly related to the survival of human groups, as well as the preservation of elements of subjective value, such as mythology and religion (MILLER, 1997).

However, with the intensification of industry and the spread of human settlements, there was a significant advance in the degradation of natural resources, which led to the start of broader movements to protect natural areas for public use, influencing currents of thought in the area to this day.

At the end of the 19th century, the consolidation of North American capitalism led to growing urbanization in the cities, but paradoxically accelerated activist movements for the defense of natural environments in the United States. The intellectual climate favorable to the creation of protected areas in the country was clearly stimulated, infecting nature defenders from many other nationalities (Australia, South Africa, and some African colonies). During this period, there were two main currents of motivation that exerted varying levels of influence (political and economic), where value and importance was attributed to nature: preservationist and conservationist. The predominance of one or other type of motivation ended up underpinning different strategies for protecting nature, spreading arrangements of protected area systems (MCCORMICK, 1992).

In view of the above, the preservationist current was strongly inspired by romantic values, focused on aesthetics and appreciation of nature, the sacredness of wild landscapes and their cultural and inspirational importance. Consequently, they advocated the protection of nature by creating areas in which no human use other than simple temporary visitation was allowed. Its fundamental values were the direct but rational, democratic and efficient use of nature. The focus of this group's attention was forest management and the multiple use of natural resources, such as fishing, hunting, among others. With regard to the creation of specially protected areas, the strategy they proposed was to defend public forests for the purposes of research and rational use, within the principles of democratization of access, efficiency in the use of resources and the production of maximum well-being, for the maximum number of people, for as long as possible (MCCORMICK, 1992).

It should be noted that in 1872, after several studies had been carried out, a law was passed designating an area of 800,000 hectares in Wyoming, in the United States, as Yellowstone National Park, the first national park in the world. Thus, the creation of the park allowed the region to remain preserved and prohibited from being colonized, occupied or sold, according to American law (MILLER, 1980, *apud* DIEGUES, 1993). The process of creating Yellowstone National Park gave impetus to the preservationist perspective, which would ensure that national parks were the only way to save scenically beautiful pieces of nature from the deleterious effects of urban-industrial development.

However, it was the creation of Yellowstone National Park in the second half of the 19th century that marked the beginning of the modern movement for nature conservation, through the creation, implementation and management of areas as we perceive and discuss them today. As such, the park could guarantee the protection of the area in the face of any exploitative activity, be it logging, mining or other, seeking to ensure the perpetuity of nature (MILANO, 2012).

It is important to remember that after the creation of Yellowstone National Park, the idea of a national park quickly conquered frontiers and, even before the turn of the 19th to the 20th century, was followed by the Royal National Park in Australia in 1872, the Banff (1885) and Glacier (1886) national parks in Canada, in 1909 Europe created a set of five national parks and in 1925 the Virunga National Park was created in Africa (Zaire) (PRESTRE, 2000). It is also worth noting that in 1933, with the World Convention for the Protection of Natural Fauna and Flora held in London, the concept of a national park was established internationally, along with a few other management categories (MILANO, 2012).

To further emphasize the importance of preserving nature, Prestre (2000) mentions that "international environmental cooperation today encompasses a variety of issues, ranging from the management of natural resources and ecosystems to the fight against pollution in various environments, the protection of biodiversity and the atmosphere and the promotion of public health." For the author, human beings need the environmental goods produced by nature, since international cooperation on the environment has become more complex to encompass all levels of international interaction.

In this sense, in 1948 the International Union for the Protection of Nature (IUCN) was established, the embryo of the current International Union for the Conservation of Nature (IUCN), an international entity that associates countries, government agencies and non-governmental organizations, with emphasis on strategies for the protection of endangered species and the most diverse important aspects in the conception of protected areas.

In Brazil, an important event of this period was the initiative, in 1876, of a proposal to create the first protected area in the country, until then, by Engineer André Rebouças who, inspired by *Yellowstone*, suggested the creation of two National Parks: one in Sete Quedas and the other on Bananal Island (MILANO, 2000). However, these units were only created many years later, despite the fact that Rebouças' work at the time already associated nature conservation with future benefits (URBAN, 1998 *apud* MORSELLO, 2001).

In 1911, the publication of the "Forest Map of Brazil" proposed the creation of a UC in Acre, but the initiative was also aborted (FUNATURA, 1989 *apud* MORSELLO *op. cit.*). Among the attempts, Decree No. 1,713 of June 14, 1937, signed by President Getúlio Vargas, created Brazil's first National Park, the Itatiaia National Park, with an original area of 11,493 ha. Later, in 1982, due to its recognition as an important area for the conservation of regional biodiversity, the Park's area was expanded to approximately 28,000 ha by Decree 87,586 (MMA, 2002).

4.4 - National System of Conservation Units

The creation and implementation of conservation units is the main policy for conserving biological diversity in Brazil and worldwide. The great challenge is to manage a system of conservation units in such a way as to guarantee their sustainability, because in addition to the countless problems that occur within each conservation unit, managers face difficulties in controlling the impacts of the activities carried out in their surroundings.

The Brazilian model of conservation units established by Law 9.985/2000, which deals with the National System of Conservation Units (SNUC), has its own characteristics that differ greatly from the North American model (DOUROJEANNI, 1997; MILANO, 2000; CABRAL and SOUZA, 2005). The SNUC is the result of almost a decade of wide-ranging debate in the National Congress, consolidating conservation units as special territorial spaces and establishing criteria and rules for the creation, implementation and management of conservation units.

Biodiversity conservation is not the only objective of the SNUC, since it also includes the protection of watersheds, natural and little altered landscapes of outstanding scenic beauty, conservation of historical, archaeological and cultural sites, restoration of degraded environments, protection of natural resources necessary for the subsistence of traditional populations, respecting and valuing their knowledge and culture, as well as the promotion of scientific research, recreation and ecological tourism (BRASIL, 2000, art. 4).

Because conservation units fulfill very different objectives, the SNUC divides them into two groups with specific characteristics: Full Protection Units and Sustainable Use Units (BRASIL, 2000, art. 7), which can be federal, state or municipal (BRASIL, 2000, art. 3).

The basic objective of the Integral Protection Units is to preserve nature, with only indirect use of its natural resources permitted (BRASIL, 2000, art. 7 § 1). Examples of categories of Integral Protection Units are I - Ecological Station; II - Biological Reserve; III - National Park; IV - Natural Monument; V - Wildlife Refuge.

I - Ecological Station: aims to preserve nature and carry out scientific research. The land is in public ownership and domain, and private areas within its boundaries must be expropriated. Public visitation is prohibited, except for educational purposes. Scientific research depends on prior authorization from the body responsible for administering the unit and is subject to the conditions and restrictions established by the latter, in an area corresponding to a maximum of three percent of the total extension of the unit and up to a limit of one thousand five hundred hectares (BRASIL, 2000, art. 9).

II - Biological Reserve: its objective is the integral preservation of the biota and other natural attributes existing within its boundaries, without direct human interference or environmental modifications, with the exception of measures to recover its altered ecosystems and the management actions necessary to recover and preserve the natural balance, biological diversity and natural ecological processes. The land is in public ownership and domain, and private areas within its boundaries must be expropriated. Public visitation is prohibited, except for educational purposes. Scientific research depends on prior authorization from the body responsible for administering the unit and is subject to the conditions and restrictions established by the latter (BRASIL, 2000, art. 10).

III - National Park: its basic objective is the preservation of natural ecosystems of great ecological importance and scenic beauty, enabling scientific research and the development of environmental education and interpretation activities, recreation in contact with nature and ecological tourism. The land is in public ownership and domain, and private areas within its boundaries must be expropriated. Public visitation is permitted, but is subject to the rules and restrictions established in the unit's Management Plan. Scientific research depends on prior authorization from the body responsible for administering the unit and is subject to the conditions and restrictions established by the latter, as well as those provided for in regulations. Units in this category, when created by the State or Municipality, will be called State Park and Municipal Natural Park respectively (BRASIL, 2000, art. 11).

IV - Natural Monument: its basic objective is to preserve rare, unique natural sites or sites of great scenic beauty. It can be made up of private areas, as long as it is possible to reconcile the objectives of the unit with the use of the land and natural resources of the site by the owners. If there is incompatibility or if the owner does not agree to the conditions proposed by the body responsible for managing the unit, the area must be expropriated. Public visitation is subject to the conditions and restrictions established in the unit's Management Plan (BRASIL, 2000, art. 12).

V - Wildlife Refuge: aims to protect natural environments where conditions are ensured for the existence or reproduction of species or communities of local flora and resident or migratory fauna. It can be made up of private areas, as long as it is possible to reconcile the objectives of the unit with the use of the land and natural resources of the site by the owners. If there is incompatibility or if the owner does not agree to the conditions proposed by the body responsible for managing the unit, the area must be expropriated. Public visitation is subject to the rules and restrictions established in the unit's Management Plan. Scientific research depends on prior authorization from the body responsible for administering the unit and is subject to the conditions and restrictions established by the latter, as well as those provided for in regulations (BRASIL, 2000, art. 13).

Sustainable Use Units have the basic objective of making nature conservation compatible with the sustainable use of part of its natural resources (BRASIL, 2000, art. 7 § 2). The following categories make up the group of Sustainable Use Units: I - Environmental Protection Area; II - Area of Relevant Ecological Interest;

III - National Forest; IV - Extractive Reserve; V - Fauna Reserve; VI - Sustainable Development Reserve; and VII - Private Natural Heritage Reserve.

I - Environmental Protection Area: a generally extensive area, with a certain degree of human occupation, endowed with abiotic, biotic, aesthetic or cultural attributes that are especially important for the quality of life and well-being of human populations, and whose basic objectives are to protect biological diversity, discipline the occupation process and ensure the sustainable use of natural resources. It can be made up of public or private land (BRASIL, 2000, art. 15).

II - Area of Relevant Ecological Interest: an area generally small in size, with little or no human occupation, with extraordinary natural characteristics or which is home to rare specimens of regional biota, with the aim of maintaining natural ecosystems of regional or local importance and regulating the permissible use of these areas, so as to make it compatible with nature conservation objectives. It can be made up of public or private land. Rules and restrictions can be established for the use of private property located in an Area of Relevant Ecological Interest (BRASIL, 2000, art. 16).

III - National Forest: is an area with forest cover of predominantly native species and its basic objective is the sustainable multiple use of forest resources and scientific research, with an emphasis on methods for the sustainable exploitation of native forests. The land is in public ownership and domain, and private areas within its boundaries must be expropriated. However, the traditional populations that inhabited the area when it was created are allowed to remain, in accordance with the provisions of the regulations and the unit's Management Plan. Public visitation is permitted, subject to the rules established for the management of the unit by the body responsible for its administration. Research is permitted and encouraged, subject to prior authorization from the body responsible for administering the unit, the conditions and restrictions established by it and those provided for in regulations. The National Forest will have an Advisory Council, chaired by the body responsible for its administration and made up of representatives of public bodies, civil society organizations and, where appropriate, traditional resident populations. The unit of this category, when created by the State or Municipality, will be called, respectively, State Forest and Municipal Forest (BRASIL, 2000, art. 17).

IV - Extractivist Reserve: an area used by traditional extractivist populations, whose subsistence is based on extractivism and, complementarily, on subsistence agriculture and small animal husbandry, and whose basic objectives are to protect the livelihoods and culture of these populations, and to ensure the sustainable use of the unit's natural resources. It is public domain, with use granted to traditional extractivist populations, and private areas within its boundaries must be expropriated. The Extractive Reserve will be managed by a Deliberative Council, chaired by the body responsible for its administration and made up of representatives of public bodies, civil society organizations and traditional populations living in the area, as provided for in regulations and in the act creating the unit. Public visitation is permitted, provided it is compatible with local interests and in accordance with the provisions of the area's Management Plan. Scientific research is permitted and encouraged, subject to prior authorization from the body responsible for administering the unit. The unit's Management Plan will be approved by its Deliberative Council. The exploitation of mineral resources and amateur or professional hunting are prohibited. Commercial exploitation of timber resources will only be permitted on a sustainable basis and in special situations that complement the other activities carried out in the Extractive Reserve, in accordance with the provisions of the regulations and the unit's Management Plan (BRASIL, 2000, art. 18).

V - Fauna Reserve: a natural area with animal populations of native species, terrestrial or aquatic, resident or migratory, suitable for technical and scientific studies on the sustainable economic management of wildlife resources. It is publicly owned and controlled, and private areas within its boundaries must be expropriated. Public visitation may be allowed, as long as it is compatible with the management of the unit. Hunting is prohibited. The marketing of products and by-products resulting from research will comply with the provisions of the laws on fauna and regulations (BRASIL, 2000, art. 19).

VI - Sustainable Development Reserve: a natural area that is home to traditional peoples, whose existence is based on sustainable systems for exploiting natural resources, developed over generations and adapted to local ecological conditions, and which play a fundamental role in protecting nature and maintaining biological diversity. Its basic objective is to preserve nature and, at the same time, ensure the conditions and means necessary for the reproduction and improvement of the ways of life and quality of life and exploitation of natural resources of traditional populations, as well as to value, conserve and improve the knowledge and techniques of environmental management developed by these populations. It is a public domain, and private areas within its boundaries must be expropriated when necessary, in accordance with the law. It will be managed by a Deliberative Council, chaired by the body responsible for its administration and made up of representatives of public bodies, civil society organizations and traditional populations living in the area, as provided for in regulations and in the act creating the unit (BRASIL, 2000, art. 20).

VII - Private **Natural Heritage Reserve:** a private area set aside in perpetuity with the aim of conserving biological diversity. There is a term of commitment signed before the environmental agency, which will verify the existence of public interest, and it will be recorded in the margin of the inscription in the Public Property Registry. Scientific research and visits can only be allowed for tourist, recreational and educational purposes. Whenever possible and appropriate, the SNUC's member bodies will provide technical and scientific guidance to the owner of a Private Natural Heritage Reserve in order to draw up a Management or Protection and Management Plan for the unit (BRASIL, 2000, art. 21).

Conservation units must have a Management Plan (BRASIL, 2000, art. 27).

4.5 - Management Plan

According to the SNUC (BRASIL, 2000, art. 2, XVII), a Management Plan is a "technical document which, based on the general objectives of a Conservation Unit, establishes its zoning and the rules that should govern the use of the area and the management of natural resources, including the implementation of the physical structures necessary for the management of the Unit".

The Management Plan supports the managers of the conservation unit in fulfilling the objectives established at the time of its creation, as it defines specific management objectives and guidelines for its development. This technical document establishes the differentiation and intensity of use by means of zoning, with a view to protecting its natural and cultural resources, highlighting the representativeness of the conservation unit in the SNUC in relation to the attributes that enhance its resources. Also present, where appropriate, are specific norms and actions aimed at making the presence of resident populations compatible with the objectives of the conservation unit, until it is possible to compensate them and relocate them (IBAMA, 2002).

In addition to the territory of the conservation unit, there are also specific rules governing the occupation and use of resources in the buffer zone and ecological corridors. These measures aim not only to protect the conservation unit, but also to promote the socio-economic integration of the surrounding communities with it.

Thus, it can be said that the Management Plan is a guiding instrument for the conservation unit, and is used on a recurring basis to support technical analysis and strategic decisions within the scope of its management. It is important to note that FLONA MX does not yet have a Management Plan, although it has already started a preliminary text for its preparation which, however, still needs to be revised and updated, as guided by the methodological roadmap. This shows that the challenge of managing FLONA MX is even greater insofar as the conservation unit does not have its own "instruction manual", leaving managers without proper support when making decisions.

The lack of a Management Plan for FLONA MX is mainly due to the high cost of outsourced services for its execution, so this Conservation Unit recently sent the Draft Proposal for the Preparation of a Management Plan to be evaluated by the Environmental Compensation Chamber - CCA of the State Secretariat for the Environment - SEA. The intention is to use part of the funds from the Rio de Janeiro State Environmental Compensation Fund to hire a company specialized in carrying out the Management Plan. It should be noted here that the Draft Proposal for the Preparation of the MX FLONA Management Plan has already been evaluated and approved by SEA technicians, presented and approved by the CCA Collegiate Body, and is awaiting a call for tenders, in accordance with the legal procedural procedures in force. The amount requested by FLONA MX to carry out the Management Plan Project is R$350,000.00 (three hundred and fifty thousand reais).

4.6 - Zoning

In the global context, zoning appeared in the 1970s in protected areas as a proposed management tool. The American Kenton Miller made a name for himself with his 1980 publication "Planning National Parks for Ecodevelopment in Latin America".

> The proposed concept consolidated zoning as the spatial distribution of different management actions, such as: "Park planning zoning differs in that it is designed to prescribe management activities for specific zones. These zones do not describe what is to be found in relation to natural resources, but rather prescribe how these resources will be located and used" (MILLER, 1980, *apud*; ELAP: 10).

Therefore, zoning is a land-use planning tool used as a resource to achieve better results in the management of the conservation unit, as it establishes different uses for each zone, according to its pre-determined objectives. Thus, zoning aims to give greater protection to natural resources insofar as each zone will be managed according to the rules established for them. By using zoning, it is possible to reconcile different uses for the same conservation unit, allowing all its objectives to be achieved in a harmonious and effective manner (BRASIL, 2000, art. 2, XVI).

In Brazil, the Regulations for Brazilian National Parks, promulgated by Decree No. 84.017 of September 21, 1979, established zoning as a planning strategy for conservation units by defining management plans. However, the approval of the SNUC opened up the possibility of including new zones that were not provided for in the National Parks Regulations, Decree 84.017/79.

The following is a list of all the zones that can be considered when zoning the internal area of a forest-type conservation unit

according to the Methodological Roadmap for Preparing Management Plans for National Forests (IBAMA, 2009):

I - Preserved Zone - where the primitiveness of nature remains as unspoiled as possible, with no human alterations tolerated. It represents the highest degree of preservation. The basic objective of management is preservation, guaranteeing natural evolution.

II - Primitive Zone - where there has been little or minimal human intervention, containing species of flora and fauna or natural phenomena of great scientific value. It should have transitional characteristics between the Preservation Zone and the Management Zone. The general aim of management is to preserve the natural environment and at the same time facilitate scientific research and environmental education activities while allowing primitive forms of recreation.

III - Community Sustainable Forest Management Zone - made up mostly of natural areas, with the possibility of some human alterations. It is characterized as a transition between the Primitive Zone and the Zones of greater intensity of use. In this zone, the needs of the traditional/local population existing in or around the Conservation Unit will be met. The aim of management is to maintain a natural environment with minimal human impact through the exploitation of forest resources, both timber and non-timber. The aim is to guarantee the integration of the National Forest into the social and economic life of the population surrounding the Unit, as recommended by the SNUC. In addition, public access can be offered with facilities for educational and recreational purposes, as well as the management of native fauna.

IV - Sustainable Forest Management Zone - comprises areas of native or planted forest with economic potential for the sustainable management of forest resources. Its objectives are: sustainable multiple use of forest resources, generation of technology and forest management models. Research, environmental education and interpretation activities are also permitted.

V - Public Use Zone - made up of natural areas or those altered by man. The environment is kept as close to natural as possible and should include: visitor centers, museums, other facilities and services. The general management objective is to facilitate intensive recreation and environmental education in harmony with the environment.

VI - Historical-Cultural Zone - where samples of historical/cultural or archaeological/paleontological heritage are found, which will be preserved, studied, restored and interpreted for the public, serving research, education and scientific use. The general aim of management is to protect historical or archaeological sites in harmony with the environment.

VII - Recovery/Restoration Zone - contains areas that have been considerably anthropized. This is a temporary zone, because once it has been restored, it will be incorporated back into one of the permanent zones. Introduced exotic species must be removed and restoration must be natural or naturally induced. The general management objective is to stop the degradation of the resources or to restore the area. This Zone allows public use for education only.

VIII - Special Use Zone - contains the areas necessary for the administration, maintenance and services of the Conservation Unit, including housing, workshops and others. These areas will be chosen and controlled so as not to conflict with their natural character and should be located, whenever possible, on the outskirts of the Conservation Unit. The general management objective is to minimize the impact of the installation of structures or the effects of works on the natural or cultural environment of the Unit.

IX - Population Zone - This comprises the homes of the Traditional Populations living within the National Forest, including the spaces and land use necessary for the reproduction of their way of life. The general management objective is to reconcile the conservation of natural resources with the needs of these populations. Visiting, environmental education and interpretation activities can only be carried out in agreement with the community.

X - Conflicting Use Zone - spaces whose uses and purposes were established before the creation of the conservation unit and conflict with the objectives of the protected area. They are occupied by public utility projects such as gas pipelines, oil pipelines, transmission lines, antennas, water catchment, dams, roads, optical cables and others. Their management objective is to contemporize the existing situation, establishing procedures to minimize impacts on the conservation units.

XI - Buffer Zone - The surroundings of a conservation unit, where human activities are subject to specific rules and restrictions, with the aim of minimizing negative impacts on the unit (BRASIL, 2000, art. 2, XVIII).

4.7 - Buffer Zone

Irregular urban growth and the various socio-economic activities carried out - from predatory extraction to hunting, logging to mining, agriculture to industry and tourism - can have a negative impact on the ecosystems that are to be protected within the unit.

In order to minimize or even avoid environmental impacts on conservation units, buffer zones have been defined, with the aim of establishing rules for activities carried out around the units, acting as buffer areas. "The buffer zone is not part of the conservation unit, nor does it prevent the development of economic activities next to it, nor does it require the expropriation of land. But the buffer zone is essential for the management of the CU" (GANEM, 2015, p.3).

The buffer zone allows the unit's manager to define zoning around the unit's boundaries, establishing control measures that are negotiated with the local populations, who will be guaranteed effective participation in the creation, implementation and management of the conservation units (SNUC Art. 5, III).

> However, considering that the delimitation and management of the ZA, in certain situations, involves major conflicts of interest between conservation and the use of natural resources by the neighboring population and other economic agents, the ideal is for it to be defined at the very moment of the creation of the UC, which is already preceded by technical studies and public consultations. It is true that the delimitation of the UC, in this case, may involve a more complex process of negotiations prior to the creation of the unit, but, on the other hand,

> it will bring greater legal certainty to the future management of the area (GANEM, 2015, p.7).

According to the SNUC Law, all conservation units, with the exception of Environmental Protection Areas (APA) and Private Natural Heritage Reserves (RPPN), must have buffer zones (art. 25, caput), with their boundaries and land use and occupation rules being defined in the act creating the unit or later (art. 25, § 2), and may even occur when the unit's management plan is established. The buffer zone can only be defined after the management plan has been approved, because it must cover the area itself (BRASIL, 2000, art. 27, § 1).

However, it should be borne in mind that the Snuc Law does not define specific criteria for delimiting the buffer zone.

> "There is no way to define, in law, general criteria for delimiting the buffer zone, not even for defining the PAs themselves. These limits must be analyzed on a case-by-case basis, always based on technical studies that guide the managing body on the ecological aspects and socio-economic activities present in the area" (GANEM, 2015, p.8).

Technical documents such as the Methodological Roadmap for Planning a National Park, Biological Reserve and Ecological Station (IBAMA, 2002) and the Methodological Roadmap for Preparing Management Plans for National Forests (ICMBio, 2009) provide guidance on important issues to be considered when delimiting the buffer zone, such as the inclusion of areas that could affect the integrity of the conservation unit, for example rivers, preserved areas, sensitive areas subject to erosion, as well as areas at risk of urban expansion. Criteria for exclusion from the buffer zone should also be considered, such as the presence of established urban areas or even areas that have not been urbanized, but which have already been defined in master plans as being subject to urban expansion.

According to ICMBio's Normative Instruction - IN 01/2007, the Buffer Zone "must be delimited taking into account regional socio-environmental characteristics, existing activities and potential impacts on the Unit. Norms for its use and occupation and restrictions on impacting activities must be established" (Art. 6, V). So, even if it doesn't belong to the conservation unit, the buffer zone is under the responsibility of the manager. It is even up to the manager to authorize the licensing of undertakings when their activities affect the conservation unit or its buffer zone, even if they do not belong to the Integral Protection Group (Art.36 § 3).

Conama Resolution 13/1990, which predated the Snuc Law, already stipulated that the body responsible for the Conservation Unit, together with the licensing and environmental bodies, defined the activities that could affect the biota of the Conservation Unit (Art. 1); and that in the areas surrounding the Conservation Units, within a radius of ten kilometers, any activity that could affect the biota had to be licensed by the competent environmental body (Art. 2).

However, Conama Resolution 13/1990 was revoked by Conama Resolution 428/2010, which stipulates that the licensing of undertakings with a significant environmental impact that may affect Conservation Units or their Buffer Zones, as deemed necessary by the licensing environmental agency, based on an Environmental Impact Study and respective Environmental Impact Report (EIA/RIMA), can only be granted after authorization from the body responsible for managing the Conservation Unit or, in the case of Private Natural Heritage Reserves (RPPN), by the body responsible for creating them (Art.1It also states that for a period of 5 years, counting from the publication of Resolution 473/2015, the licensing of undertakings with a significant environmental impact located within a 3,000 meter strip from the boundary of the Conservation Unit, whose Buffer Zone has not been established, will be subject to the prescribed procedure, with the exception of RPPNs, Environmental Protection Areas (APAs) and Consolidated Urban Areas (Art.1 § 2).

Thus, it can be seen that the legislation has been updated, bringing it closer to the reality of many conservation units, including FLONA MX, which does not yet have a management plan and which is under pressure from various impacting activities and enterprises and a consolidated urban area. In this sense, the environmental zoning proposal for FLONA MX will promote better knowledge of the conservation unit and its surroundings, strengthening its potential so that in the future a Management Plan can be drawn up that can mitigate or at least reduce environmental conflicts in its territory and the surrounding area. Only in this way will the conservation unit comply with the SNUC Law and be able to access or apply for funds from other sources, moving closer to the government's goals of modern political-environmental management.

Chapter 5

5 - MATERIALS AND METHODS

In order to propose environmental zoning that justifies the maintenance of FLONA MX as a conservation unit, it is necessary to list and locate the potential and weaknesses, as well as the environmental conflicts that affect FLONA MX. To do this, we had to look for

- abiotic variables such as altitude, slope, relief orientation, solar radiation and hydrography;
- biotic variables, such as vegetation (native and exotic) and habitats of endemic species threatened with extinction (birds and fish);
- land use and cover and its conflicts with environmental legislation;
- urban infrastructure, such as roads and buildings.
- legislation, alphanumeric data and bibliography, such as the Methodological Roadmap for Preparing a Management Plan for National Forests (ICMBio), the Forest Inventory and endemic species, the Draft Management Plan and dissertations on the study area.

This information was georeferenced to be worked on in a Geographic Information Systems environment, using the open-source Quantum GIS software, which used satellite image classification, NDVI, vectorization and geoprocessing tools.

The work was divided into the following stages:

1. Collecting data from public and private sources:

Information was collected from IBGE websites, where it was possible to extract orthophotos of the region, shapes and statistical data. We consulted the Seropédica City Hall, which provided environmental data on the region; the UFRRJ, especially the scientific papers produced at the institution; and the management of FLONA MX, which provided various data and access to documents. The satellite images were taken from the INPE and GLOVIS websites.

2. Vectorization of environmental features through photointerpretation of orthophotos and satellite images

Vectorization of geographical features such as trails, paths, lakes, buildings and some streams using orthophotos and satellite images was carried out in the Quantum GIS environment.

3. Generation of a Terrestrial Surface Model (TSM)

The abiotic variables of altitude, slope, relief orientation and solar radiation were extracted from the elevation model generated using data from Terra satellite images, from its Aster imaging instrument, with the Visible (Blue, Green and Red) and Near Infrared spectrum bands; and from elevation data provided by the city hall and FLONA MX. The MDT was generated in Quantum GIS using geoprocessing tools.

4. Image classification and generation of land use and land cover shapes

Satellite images (Rapideye) were classified in QGIS using geoprocessing tools to generate land use and land cover data.

5. Generation of Vegetation Indices using satellite images

Satellite images (Rapideye and Landsat) were classified in QGIS using geoprocessing tools to generate vegetation indices.

6. Generation of the Environmental Conflict Raster

Cross-referencing data on permanent protection areas (APP) and land use and cover to extract information on environmental conflicts in the conservation unit.

7. Making thematic maps

The following maps were made using QGIS layout tools: map of the Buffer Zone with land use and cover; map of the location of FLONA MX; map of the Use and Cover of the CU; map of Environmental Zoning (APPs); map of Environmental Conflicts; map of the CU's Vegetation and map of Environmental Zoning with the final result of the analysis.

8. Spatial analysis and mapping of environmental fragility

This was done using geoprocessing techniques (Weigthed overlayer) and abiotic data on altitude, slope, relief orientation, radiation; biotic data on vegetation and biodiversity (area covered by endangered species - Ra and Fish); and anthropic data on land use and occupation and conflicting use.

In this way, the areas with the greatest environmental fragility were generated, in which numerical values were reclassified to meet a qualitative level of vulnerability of the natural vegetation.

Chapter 6

6 - RESULTS AND DISCUSSION

- Mapping Potential

Two maps were drawn up to analyze the potential. The first is a map of land use and cover, showing the layout of existing buildings, the classification of vegetation, paths and trails (Figure 20).

It can be seen in red that, internally, FLONA MX has some roads and that the buildings are located along these roads, mainly near the main entrance (BR-465), a stretch known as the main road, and on the old Rua da Floresta, currently occupied by AMRJ. It is worth noting that the long stretch of Estrada do Portao, which crosses FLONA MX up to BR-116, has no buildings.

With regard to the vegetation, it can be seen that the old eucalyptus stands are located in the southern portion of FLONA MX, along the BR-465 highway, in a strip bordered by the Drago Ditch. In the northern portion, along the BR-116 highway, on both sides, is the area occupied by recent eucalyptus trees, such as the Saint Gobain Company. Noteworthy are the patches with well-preserved vegetation and other areas in a less advanced stage of reconstruction.

On one of the boundaries of FLONA MX with Bairro Boa Esperanga, there is a long stretch of bare ground crossing the UC in the direction of BR-465, where replanting projects of native species of the Atlantic forest are currently being carried out, totaling

6.25 hectares. There are also patches of bare soil in this area that are used illegally as pasture for cattle and horses.

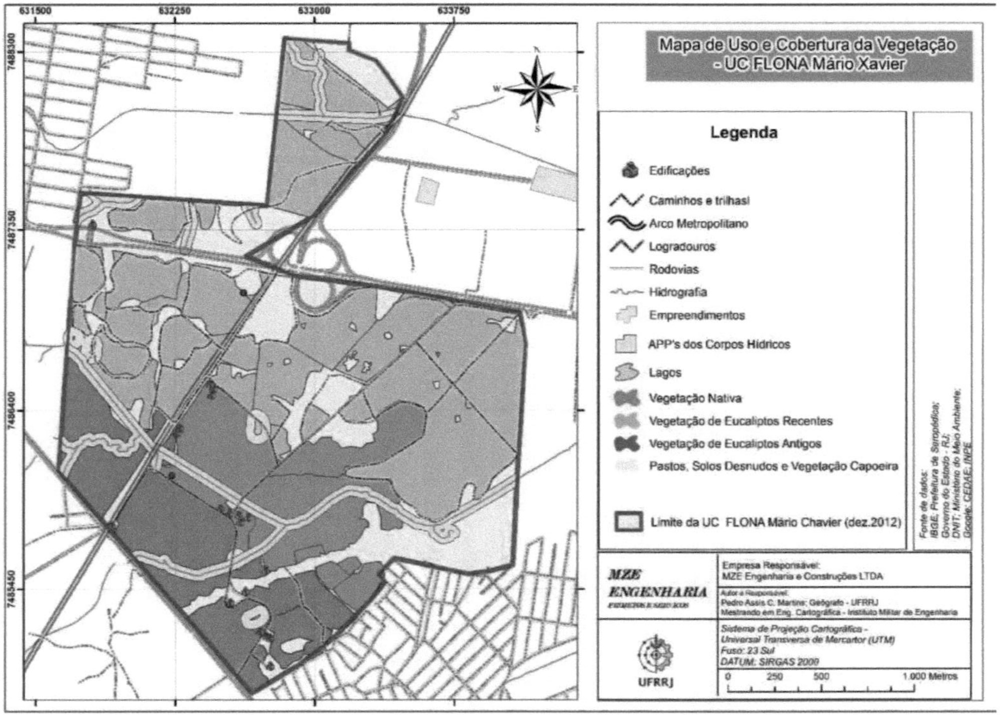

Figure 21 - Map of Vegetation Use and Cover within FLONA MX

Bare soil can also be seen near the intersection of BR-116 and BR-493, caused by the use of the area as a dismantling site for the AMRJ works.

The second map highlights the areas of occurrence of endangered species of fauna - the frog *Physealemus Soaresi* and the fish *Notolaebias minimus - which* are endemic to FLONA MX. The presence of the fauna surveyed is directly related to the existence of the pools, temporary and/or perennial, where the gene flow and reproduction of the two species occur (Figure 21).

The map shows nine points of occurrence of the species, most of which are located in the wettest areas of the old eucalyptus stands, where six points are found, two of which are located close to the boundary with the native forest, and one more point in the native forest stand itself.

In addition to these seven points, there are two more points of occurrence of endemic fauna species located in the recent eucalyptus grove, which, despite being an area with a lower density of vegetation, also has two flooded areas. One of them is to the northwest of the UC, where the presence of hills drains rainwater to the lower part of the terrain, forming a swamp. The other area, on the side of the BR-116 highway, was recently formed as a result of the construction work for the AMRJ. This flooded area was formed by the damming up of rainwater from the AMRJ's runways which, when draining into a drainage network that has not been fully completed, is thrown into FLONA MX.

Curiously, unfinished work has created a habitat for the species, but it should be remembered that this is a temporary situation, as the resumption of these drainage works will prevent the flooded area from being maintained.

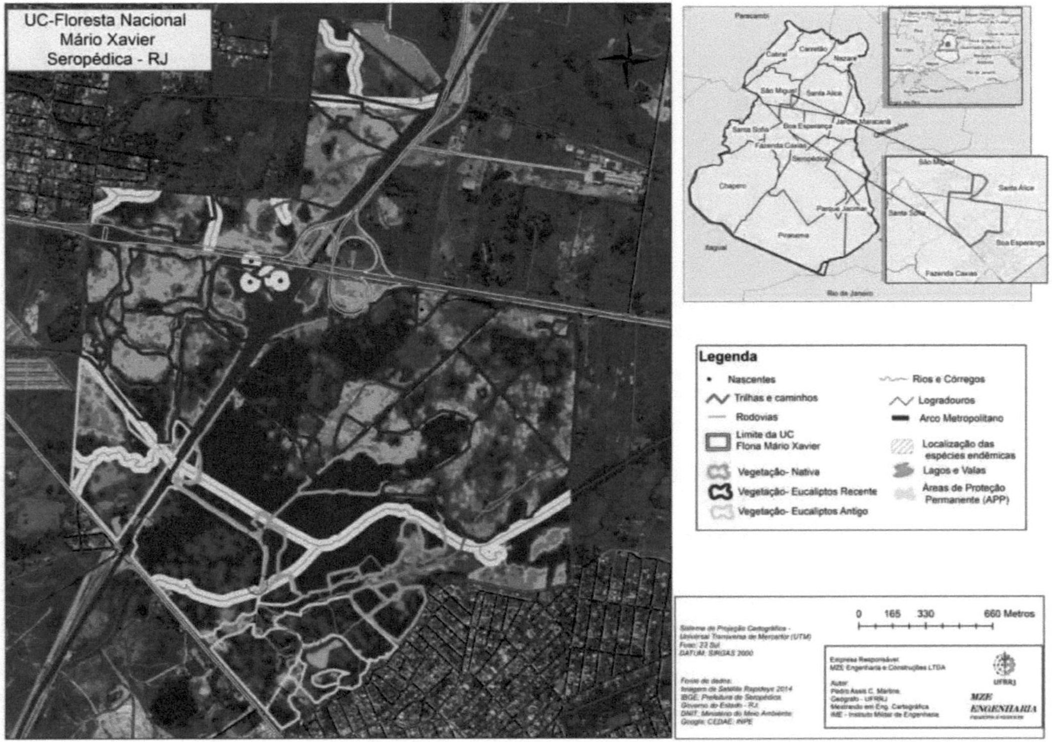

Figure 22 - Map of Vegetation Use and Cover, location of endemic species

- Conflict mapping

In order to analyze the conflicts, a mapping exercise was carried out in which some characteristics peculiar to FLONA MX are presented, as it is included as an urban Conservation Unit. For example, there is the discharge of sewage in natura into open ditches that affect FLONA MX at two points: on its border with the Boa Esperança neighborhood, to the southeast of the Conservation Unit and, in its western portion, on the border with the "Minha Casa, Minha Vida" housing development - Jardim das Acácias and Parque das Flores, where organic waste contaminates the Valao do Drago stream (Figure 22).

There are also two solid waste dumping points on the border with the Boa Esperança neighborhood, where plastic bags containing household waste have been found. Highlighted in dark yellow is the area impacted by the AMRJ, where various types of material have been found, such as: a complete sofa set, a wooden cabinet, plastic bags and bottles, dozens of truck tires that have been improperly disposed of, as well as vehicles that are involved in accidents and fall into FLONA MX (five vehicles so far).

The road used to be used as an alternative route to avoid the BR-116 toll. However, it lost its functionality with the creation of the AMRJ, which allows access to the municipality of Seropédica without having to pass through the tollbooth.

Highlighted in red are the areas used improperly as grazing land for horses and cattle, which occur on the borders with the Boa Esperança and Sao Miguel neighborhoods, where criminal fires are set to renew the weeding. To prevent this irregular practice, measures are taken to curb invasion, such as the systematic monitoring of these areas by civil servants who are constantly redoing the fences on the borders with these neighborhoods. However, on weekends, when there are no employees, these practices are carried out again.

In orange, the areas where irregular logging takes place are highlighted, especially in the area of recent eucalyptus plantations. The large number of fallen logs attracts trucks to transport this material. This is one of the most difficult practices to curb, as it usually takes place on weekends and along the BR-116 and AMRJ highways. These roads divide the unit, making internal circulation difficult and facilitating violations on the edges of FLONA MX.

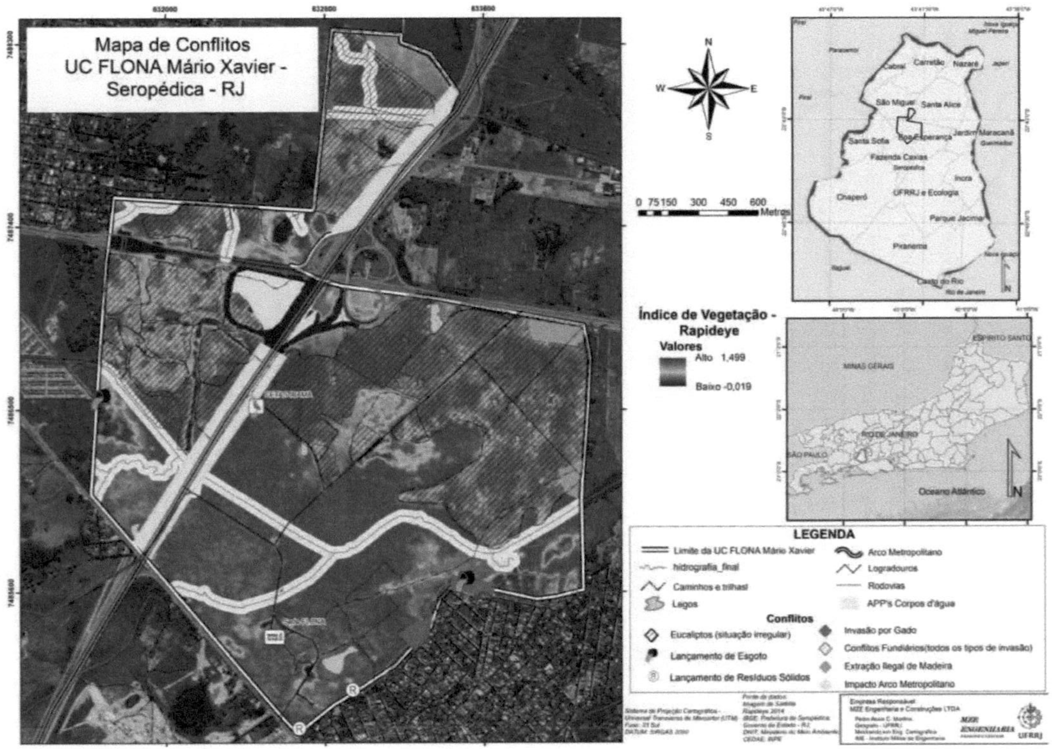

Figure 23 - Map of Conflicts in FLONA MX

With regard to FLONA MX's internal activities, the biggest conflict is the presence of recent eucalyptus plantations introduced by thc Sant-Gobain company, which have been in court for almost twenty years, as mentioned above. These eucalyptus trees represent a threat to the very existence of FLONA MX as a Conservation Unit, as they undermine its objective of preserving and conserving native species. In addition, the eucalyptus groves have been the scene of fires that have consumed almost 100 hectares in the last three years. This has prompted the Federal Public Prosecutor's Office to call in ICMBio, demanding that a permanent fire brigade be hired for FLONA MX. The ICMBio Specialized Prosecutor's Office is currently responding to this request.

Another conflict is the presence of IBAMA's Wild Animal Screening Center (CETAS). This is the only center in the state of Rio de Janeiro that receives seized animals and screens them for reintroduction into the wild. It represents a conflict because, as it receives animals from all over the state, in different health conditions, it enables the spread of zoonoses to the local fauna.

- Mapping the buffer zone

FLONA MX has no Management Plan or buffer zone, as mentioned above. In this sense, using the recommendations of Conama Resolution 428/2010, which deals with the need to maintain a strip of 3,000 meters from the boundaries of a Conservation Unit, the buffer zone was mapped (Figure 23).

According to the map, in this perimeter we see the presence of the urban fabric of the Sao Miguel, Boa Esperança and Fazenda Caxias neighborhoods, as well as some large logistics and industrial enterprises (Brasilit, Panco Alimentos, Empresa Brasileira de Agregados Minerais - EBAM), licensed by the State Agency (INEA), most of them with the consent of ICMBio.

In this way, it can be seen that the surroundings of the MX FLONA are quite transformed, including the presence of enterprises that cause significant impacts, with the characteristics of areas that could be excluded from the buffer zone even though they are within this 3,000 meter radius.

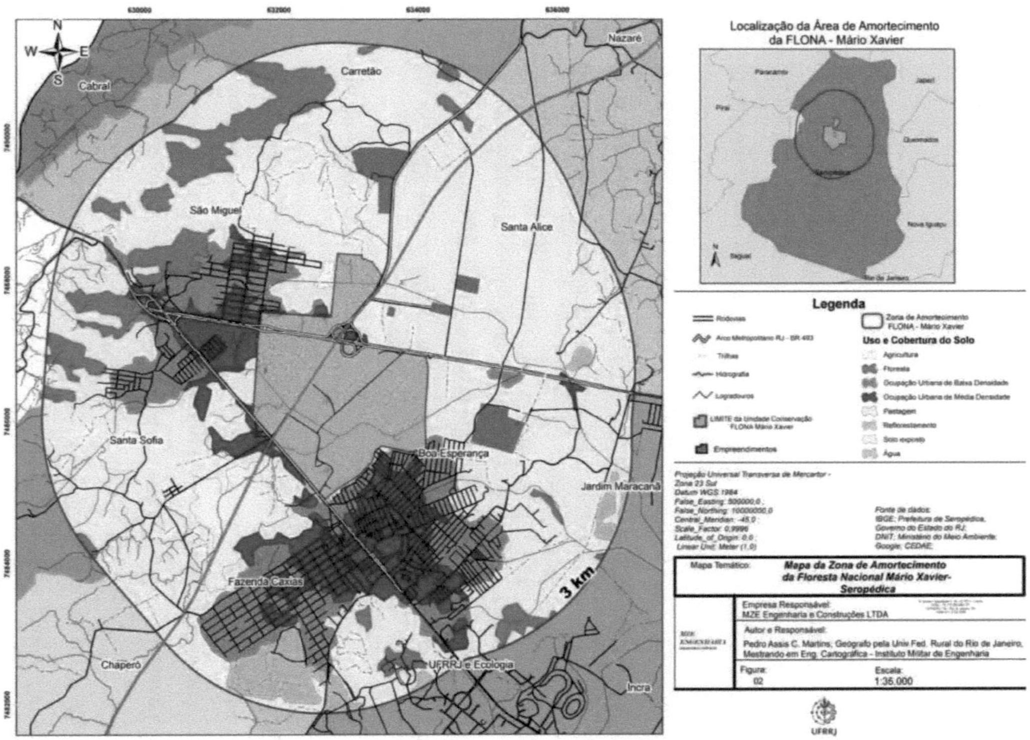

Figure 24 - Map of the FLONA MX buffer zone

It should be noted that currently, of the 327 Federal Conservation Units, 179 have a Management Plan and their buffer zones and another 98 are in the process of being drawn up or revised. In this sense, there are 50 Conservation Units that don't have a Management Plan or zoning, given that organizational and personnel limitations restrict ICMBio's potential to serve all the registered Conservation Units.

FLONA MX presented a project to draw up the Management Plan, with funds from the Environmental Compensation Chamber, which was approved at the meeting on April 11, 2017, and is expected to be completed in around two (2) years, as soon as the funds are actually released.

- Zoning mapping

For this mapping, zoning was established with different content and rules of use, taking into account the data and information from previous mappings, as well as from the Forest Cover Inventory and Assessment (AMORIM, 2007). The zones followed the lines of the polygons formed by the homogeneous plots, remnants of the history of this area used for planting seedlings of native species and for the production of various forest essences (Figure 24).

Area I, in dark green, has been defined as a Preservation Zone, where there is native Atlantic forest vegetation and therefore no human alterations are allowed. It acts as a matrix for other zones where regulated human activities are already permitted. This zone is dedicated to the integral protection of ecosystems, guaranteeing their natural evolution. Its use is characterized by a preservationist level with rules of availability for scientific research activities in a restrictive manner, when impossible to carry out in other areas of the FLONA.

In this area, although the vegetation is at a lower stage of regeneration and even needs induced intervention with the enrichment of native Atlantic forest species, the presence of the lake with the endemic species makes this area intangible. It is worth noting that this lake is the one that presents the best conditions for preservation, as the others are very close to pedestrian circulation areas, at the main entrance to FLONA MX, or in the vicinity of AMRJ.

Areas II and III, in light green, have been defined as the Primitive Zone, an area with little human intervention that contains native forest species. The aim of this zone is to create protection for the Preservation Zone. In area II there is a strong influence of native species planted at an advanced stage of natural regeneration, in a good state of conservation.

In area III there is a strong influence of planted native species, at an advanced stage of natural regeneration, however the presence of the Seedling Nursery makes this an area of permanent use for internal activities, and it can also be open to visitors. At the moment the Seedling Nursery is deactivated due to the poor structural condition of the building, but its ongoing restoration project as compensation for AMRJ's LI has already been approved and has not yet been implemented.

Areas IV, V and VI, in yellow, have been defined as Community Sustainable Forest Management Zones, consisting of natural areas with some human alterations where the needs of the population surrounding FLONA MX will be met. Religious services take place in these areas, where the main impact comes from the systematic sweeping up of fallen leaves, which removes the leaf litter substrate. The aim of these Zones is to protect the Primitive Zone, maintaining a natural environment with minimal human impact, while at the same time integrating the social life of the community into FLONA MX.

Area IV is covered by Sapucai trees in 2 X 2 meter spacings, where there has been no understory regeneration, probably due to the closed canopy preventing sunlight from penetrating. Area V is adjacent to the Sapucai trees.

Area VI is relatively degraded due to its natural, physical-biological soil and topographical characteristics, which in the past made it possible to exploit it economically by removing gravel. Today, on a smaller scale, the substrate mixture is used to produce forest seedlings in the UC's own nursery. This area is covered by secondary vegetation at an early stage of natural regeneration, where the successional process occurs very slowly, and its hilly topography means that the population has chosen it for prayers at Monte de Deus.

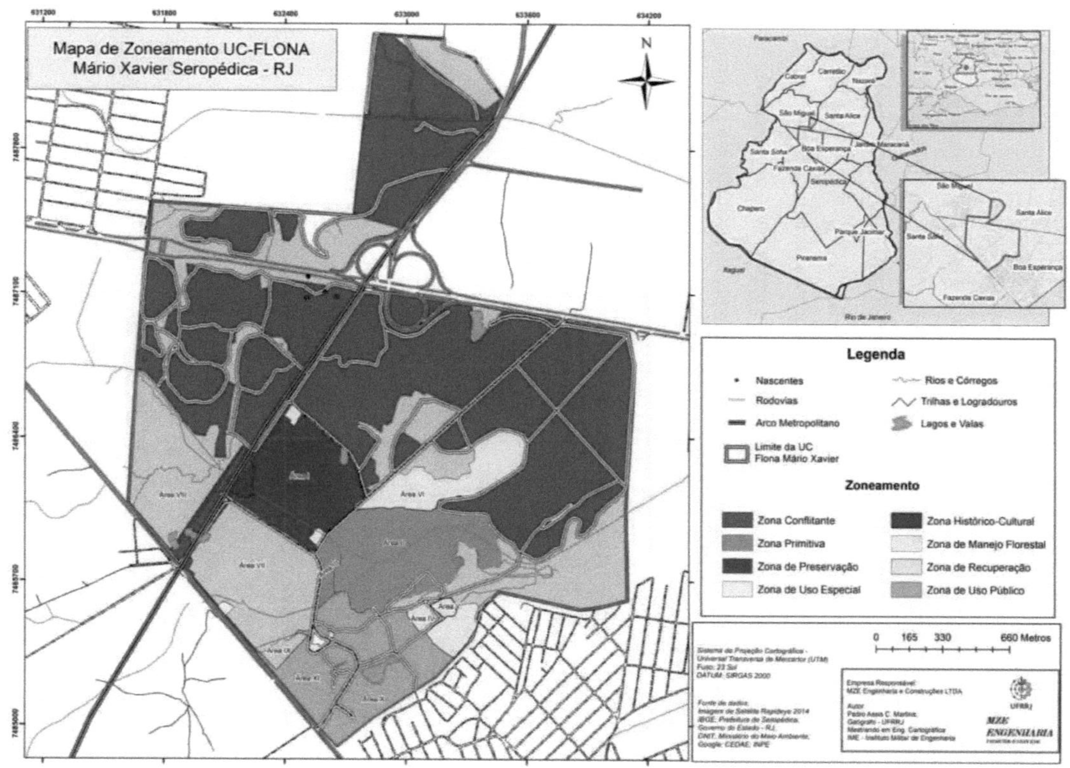

Figure 25 - Zoning proposal for FLONA MX

Areas VII, VIII and IX, outlined in green, have been defined as a Recovery Zone, as they are considerably anthropized. This is a provisional zone with the aim of halting the degradation of natural resources and restoring areas. Areas VII and VIII are occupied by exotic species, mainly eucalyptus trees, which will have to be removed in order for them to recover with the introduction of native Atlantic Forest species. The total volume of old eucalyptus trees in FLONA MX is 9,100.3 m3, and these two areas have the largest volumes, amounting to almost 50% of the Unit's total. Because they have been planted for a long time, these eucalyptus trees are well developed and belong to quality class I, which gives them good prospects for sale, provided their Management Plan is approved.

Area IX is formed by the extinct FURNAS transmission line, consisting of pastures, bare soils and capoeiras. This area has been undergoing reforestation with species native to the Atlantic Forest. There are currently around 7 hectares already planted and in the maintenance phase, and a further 6 hectares in the project analysis phase by ICMBio, awaiting approval.

In orange, the Public Use Zone was defined as an area intended for recreation and intensive use by visitors, mainly for environmental interpretation/education, recreation, leisure, research and environmental monitoring, who must use it in complete harmony with the environment. In FLONA MX, this area is made up of old eucalyptus trees and is located at the entrance to the Conservation Area, where you will find the visitors' center, the start of several interpretation trails (Trilha do Lagarto), as well as other facilities and services for public use. Public visitation will be encouraged and monitored, so as to allow contemplative viewing of the landscape.

In purple, the Historic-Cultural Zone has been defined, which is characterized as being where historical, cultural and archaeological samples are found, which will be conserved and interpreted for the public. The activities allowed are scientific research, environmental education and interpretation, where the protection of historical and archaeological sites must be in harmony with the environment. In FLONA MX, this area was home to the old Santa Cruz Farm Pottery, which is now in ruins. During the AMRJ construction work, indigenous artifacts were found in the vicinity of the pottery, which gives this part of the conservation area its distinctive character.

In light pink, the Special Use Zone was defined, which is characterized as containing the areas necessary for the administration, maintenance and services of FLONA MX. In order to make the structures and works necessary for the management of the Conservation Unit compatible with the natural environment, minimizing their impacts. Some rules of use must be prioritized, such as maintaining building standardization, environmental aesthetics and harmonious viability of the various accesses and armaments existing in the CU. These areas are represented by the administration buildings, such as the headquarters, workshop, carpentry, sawmill, seedbeds and residential properties.

In red, it has been defined as a Conflict Use Zone, as these are areas located within a Conservation Unit whose uses and purposes conflict with the conservation objectives of the protected area. These are areas occupied by public utility projects, represented by the passage of the Presidente Dutra BR-116 highway, as well as the entire stretch of the AMRJ that cuts through FLONA MX, but also areas where the vegetation cover is made up of recent eucalyptus reforestation. These species represent conflicts, both in terms of their legal institutional aspect and the introduction of monoculture in an area that should be destined for the reintroduction of native Atlantic forest species.

From a legal point of view, the area has been under litigation since 2007, and for this reason the eucalyptus trees are not allowed to be managed. As they occupy a very large area, around 250 hectares, more than half of the area of FLONA MX, they weaken the proposals for environmental protection and the very existence of the Conservation Unit. In relation to the monoculture of eucalyptus species, it is highly flammable, especially during the dry season, when the biggest fires occur in the Unit, initiated by different external factors. This Zone will be constantly controlled through specific monitoring and duly supervised, with the implementation of measures aimed at minimizing the environmental impacts generated by the activity.

CONCLUSION

The study of the FLONA MX protected area analyzed in this dissertation was based on the Ecological Zoning proposal drawn up through internal and external data surveys. The analysis was based on the existing legal framework for protected areas in Brazil and experiences abroad, as well as relevant bibliographical research on the subject.

In the course of the research we tried to highlight the main advances made in the protection and conservation of conservation units, as well as pointing out the potential, weaknesses and environmental conflicts that exist both in the historical base and in the analysis of the data generated in institutional management, which would be corroborating in the construction of a Management Plan, as an important management tool for the Conservation Unit.

It is important to note that there was a Draft Management Plan drawn up by FLONA MX staff, but this document was based on the Mapping, Inventory and Assessment of Forest Cover in the Mário Xavier National Forest (AMORIM, 2007), as well as the Technical Notes drawn up by researchers from the National Center for Research and Conservation of Continental Fish (CEPTA) and the National Center for Research and Conservation of Reptiles and Amphibians (RAN). This inventory is a very rich document that describes the plant species and their stands, taking measurements of the spread of individuals, the number of individuals and the areas occupied by these stands. The Technical Notes are also very rich and well-instrumented with regard to the preservation and conservation of endemic species (ray and cloud fish). However, these documents were strictly focused on the interior of the conservation unit and did not include the occupation of the surroundings and the relationship developed by the population with the FLONA MX area, which often generates conflicts that need to be understood and considered in the process of building the Management Plan.

Therefore, the proposal presented here used the aforementioned documents, but mainly ICMBio's Methodological Roadmap for Preparing National Forest Management Plans, as well as field activities and satellite image analysis to consider the surroundings, which generated different results in terms of the location and content of the zones established in this research work. One example is not having considered all the lakes and their surroundings as intangible areas, even when there was the presence of both ra and cloud fish. This is due to the fact that some lakes are located near the entrance to FLONA MX or on the internal circulation routes. Considering them intangible would make some of the activities currently carried out unfeasible, which would imply the prospect of re-categorizing this Conservation Unit.

There is currently a proposal to re-categorize this Conservation Unit, which would no longer be a National Forest but a Wildlife Refuge, an Integral Protection unit - designed to protect natural environments, in which the aim is to ensure conditions for the existence or reproduction of species or communities of local flora and fauna - in order to achieve a higher level of protection for endemic species. This re-categorization would imply the suspension of the use of this UC by the surrounding community, which would require much more intense monitoring and inspection, as conflicts would only tend to increase.

Thus, with the implementation of a Management Plan - an important management tool - after it has been approved by the Advisory Council of this Conservation Unit, it is hoped that FLONA MX will be better prepared to receive the visiting population, as well as to deal with events that today are treated as adversities. The creation of a Management Plan for a Conservation Unit such as a National Forest provides for the allocation of areas for the sustainable use of forest resources, whether timber or not, with a view to meeting the needs of the surrounding population while maintaining a minimum impact. In the case of FLONA MX, the surrounding population has used the Conservation Unit to meet a specific need to hold religious services, mainly Protestant. The municipality of Seropédica has the highest percentage of Protestant population in the state of Rio de Janeiro. In this zoning proposal, areas are being set aside for this purpose, which today, due to the lack of a Management Plan, is considered a conflicting activity.

We will only be able to tell if the National Forest category is suitable for a Conservation Unit where there are two cataloged endemic species threatened with extinction. We believe that a well-designed Management Plan, with a zoning proposal that is suited to the needs of FLONA MX, for the preservation of species of fauna and flora, as well as suited to the needs of the surrounding population, and that is widely discussed with this population and their representatives on the advisory council, are fundamental tasks for achieving the reduction/mitigation of conflicts, even in a Sustainable Use conservation unit with endemic species threatened with extinction.

Amaral (1998), in a study on the compatibility between integral protection units and rural communities, concluded that "it is possible for traditional communities to remain in limited portions of parks, keeping most of the area of the conservation unit under an integral protection regime" (AMARAL, 1998, p. 173). However, the community must retain its traditional characteristics, since abandoning the agricultural practices of the past and adopting intensive farming practices, with the use of pesticides and chemical fertilizers, are incompatible with the objectives of a conservation unit.

In some cases, reducing the boundaries of a conservation unit or re-categorizing it is one of the ways to minimize such conflicts (BARBOSA; MUSSI, 2009).

Bibliographical references

AB'SABER, A. N. Conceptual bases and the role of knowledge in impact forecasting. In: MULLER - PLANTENBERG, G. and AB'SABER, A. N. (eds.). Impact Forecasting: Environmental Impact Assessment in the East, West and South. Experiences in Brazil, Russia and Germany. Sao Paulo: Edusp, 1994.

ACSELRAD, H. Environment and democracy. Rio de Janeiro: Ibase, 1992.

AMORIM, H. B. Mapeamento, Inventário e Avaliagao da Cobertura Florestal da Floresta Nacional Mário Xavier, Seropédica, RJ. Final Report, 2007.

BARBOSA, C. S. Recategorization of Conservation Units: the discourse of a new territoriality and social participation in the context of the Pontoes Capixabas National Park - ES. Master's dissertation. Federal University of Minas Gerais. Institute of Geosciences, 2013. 156f.

BRAZIL. Law No. 6.938, of August 31, 1981. Provides for the National Environmental Policy, its purposes and formulation and application mechanisms, and makes other provisions.

BRAZIL. Law No. 9.985, of July 18, 2000. Establishes the National System of Nature Conservation Units and makes other provisions.

BRAZIL. Lessons learned on Zoning in and around Protected Areas: teaching and learning communities in Protected Area Planning / [WWF-Brazil] / [Latin American School of Protected Areas - ELAP]. - Brasília: WWF-Brazil, 2015. 60p.

BRITO, M. C. W. Conservation Units: intentions and results. Sao Paulo: Annablume : FAPESP, 2000.230p.

CABRAL, N. and SOUZA, M. Áreas de Protedo Ambiental: Planejamento e Gestão de Paisagens Protegidas. São Carlos: RIMA, 2005. 158 p.

CARELLI, S.G. Evolugão Geológica Neógena da Planicie Costeira de Itaguaí, 2008. 191p. PhD Thesis - Postgraduate Program in Geology. Federal University of Rio de Janeiro, Institute of Geology, Rio de Janeiro.

COIMBRA FILHO, A.F. & MITTERMEIER,R.A. Conservation of Brazilian Lion Tamarins (Leontopithecus rosália). In: Primate Conservation, 2: 59-94. New York: Academic Press Inc. 1977.

CHRISTOFOLETTI, A. Modelagem de Sistemas Ambientais. São Paulo: Edgard Blucher, 1999. 236p.

CORREA, R. L. (eds). Geography: Concepts and Themes. 8 ed. Rio de Janeiro: Bertrand, 2006.

CRUZ, C.A.G. & PEIXOTO, O.L. Description of a New Species of Annual Fish from the State of Bahia, Brazil (CYPRINODONTIFORMES, RIVULIDAE). In: Revista Brasileira de Zoologia, 7(4): 637-641. S.d. 1991.

DARLAN, A. D. Socio-environmental Conflicts Arising from the Human Presence in Conservation Units: A Case Study of the São Roque Quilombola Community in the Aparados da Serra and Serra Geral National Parks. Master's dissertation. University of the Extreme South of Santa Catarina. Graduate Program in Environmental Sciences, 2010.

DIEGUES, A. C. Traditional populations in conservation units: the modern myth of untouched nature. São Paulo: CEMAR/USP/NUPAUB, 1993.

DIEGUES, A. C. The Modern Myth of Untouched Nature. Hucitec, 1996.

DOUROJEANNI. M. Procedures for sustainable development applicable to the management of municipalities, micro-regions and river basins. ECLAC Series. Environment and Development. No. 3. ECLAC-UN. 1997; 72 pg.

DUDLEY, N. et al. Towards effective protected areas systems: an action guide to implement the Convention on Biological Diversity Programme of Work on Protected Areas. Montreal: Secretariat of the Convention on Biological Diversity, 2005 (Technical series No. 18).

Environmental Impact Study - Rio de Janeiro Metropolitan Arch Implementation Project BR 493/RJ 109. Volume I and II - DER-RJ - CONCREMAT AND TCNOSOLO Consortium. June 2007.

ENGEMANN, C. Os Servos de santo Inácio a serviço do Imperador: Demografia e relações sociais entre a escravaria da Real Fazenda de Santa Cruz, RJ. (1790- 1820); Master's dissertation - Federal

University of Rio de Janeiro, Postgraduate Program in Social History, Rio de Janeiro, 2002.

EVANGELISTA, H. A. A Fusao dos Estados da Guanabara e Rio de Janeiro Segundo uma Perspectiva de Análise Geográfica. Doctoral Thesis - Federal University of Rio de Janeiro, Postgraduate Program in Geography, Rio de Janeiro, 1998.

FRANCO, J. L. A. & SCHITTINI, G. M. History of protected areas: objectives and justifications for nature protection. In GARANDA, G.A. (org); CARVALHO, E.R. (coord). Rivers and Cities - Views of History and the Environment. Goiânia: Ed. da PUC de Goiás, 2010, p. 203-227.

FUNDACIÓN FORO NACIONAL POR COLOMBIA. Conflict, Coexistence and Democracy. Colombia, 2001.

GANEM, R. S. Buffer Zones for Conservation Units. Brasília: Legislative Consultancy, 2015.

GUERRA, A.T. & GUERRA, A.J.T. Novo Dicionário Geológico - Geomofológico. Rio de Janeiro: Bertrand Brasil.1997.652p.

GOES, H.A. A Baixada de Sepetiba. Rio de Janeiro. s.d. 1942. 245p.

ICMBio. Management Councils of Federal Protected Areas: a guide for managers and councilors. 88p. Brasilia, 2ª Edition 2015.

LACORTE, R. & NJ, E, Natural History Notes some Cynolebias Species. In: Tropical Fish Hobbyist, 11: 215-226. New York: s.d. 1989.

LOUREIRO... [et al.]. Environmental education and participatory management in Conservation Units. 3 ed. revised and updated: Ibama/NEA/Rio de Janeiro, 2008.

MACHADO, A.B.M; DRUMMOND, G.M. e PÁGLIA, A.P. Livro vermelho da fauna brasileira ameaagada de extingao. Brasília, DF: MMA; Belo Horizonte, MG: Fundagao Biodiversitas, 2008.

MARQUESE, R. B. A Administragao do Trabalho Escravo nos Manuais de Fazendeiro do Brasil Império, 1830-1847. REVISTA DE HISTORIA - Portal de Revistas USP, no. 137, p. 95-11, 1997.

McCORMICK, John. Towards Paradise: The History of the Environmental Movement. Rio de Janeiro: Relume Dumará, 1992.

MILANO. M. S. Concepts, history, policies and forms of management. Protected areas / Fundo Vale - 1.ed. - Rio de Janeiro: Fundo Vale, 2012.170p.

MILANO, M. S. Myths in the Management of Conservation Units in Brazil, or The Real Threat. In: CONGRESSO BRASILEIRO DE UNIDADES DE CONSERVAÇÃO, 2., 2000, Campo Grande, MS. Proceedings... Campo Grande: 2000, v. 1.p. 11-25.

MILLER RW (1997) Urban forestry: planning and managing urban green spaces, 2nd ed. Prentice Hall, New Jersey.

MONTEIRO, C. A. F. Human-Nature Interaction in the Future of the City. Department of Geosciences. CCH/USC. *Revista Geosul*, Santa Catarina, 14, p. 7-48, 1992.

MORAES... [et al.]. Technical manual for the restoration of degraded areas in the state of Rio de Janeiro - Rio de Janeiro: Rio de Janeiro Botanical Garden, 2013.

MORSELLO, C. Public and Private Protected Areas: Selection and Management. Sao Paulo: Annablume: Fapesp, 2001. 344p.

MORIN, Edgar. Science with Consciousness. Rio de Janeiro. Editora Bertrand Brasil, 2005.

PORTO-GONQALVES. Carlos W. Political Geography and Sustainable Development. Terra Livre Magazine. Geografía, Política e Cidadania, Sao Paulo, AGB, No. 11-12, p. 976, 1992.

REBOUQAS, A. C.; BRAGA, B.; TUNDISI, J. G. (eds.). Fresh waters in Brazil: ecological capital, use and conservation. Sao Paulo: Escrituras Editora, 3 Edigao, 2006.

Monitoring Report on the Populations of the Ra Physalaemus soaresi and the Annual Fishes Nometolebias whitei and Notolebias mínimus in the Mário Xavier National Forest, Seropédica Municipality, Rio de Janeiro. Skill Engenharia. January 2016; 88 pg.

ROCHA, A. S. The ideal representations of a territory: economic and political dynamics, agents and

the production of meanings in the territorial appropriation of Baixada Fluminense after 1990. 242 fl. - Doctoral Thesis - Postgraduate Program in Geography UFRJ, 2014.
ROCHA, A. S.; OLIVEIRA, L. D.; BARROS, R.C. (eds.). The Geography of Seropédica: theoretical reflections and educational practices in PIBID. Nova Iguagu - RJ: Entorno, 2015. 154 p.

ROSS, Jurandyr L. S. Ecogeografia do Brasil: subsídios para planejamento ambiental. Sao Paulo: Oficina de Textos, 2006. 208p.

SANTOS, A. M. S. P. Economia, espaço e sociedade no Rio de Janeiro. Rio de Janeiro. Editora FGV. 2003.

SANTOS, Jader de O. Relationships between Environmental Fragility and Social Vulnerability in Risk Susceptibility. Revista Mercator, Fortaleza, v. 14, n. 2, p. 75-90, mai./ago. 2015.

SANTOS, Luiz, A, F. Mário Xavier National Forest: A Proposal for Environmental Planning. Master's dissertation - Federal Rural University of Rio de Janeiro, Postgraduate Program in Environmental and Forestry Sciences, Forestry Institute, Rio de Janeiro, 1999.

SAQUET, M. A. and SPOSITO, E. S. (eds). Territories and Territorialities: Theories, Processes and Conflicts . 1.ed. Sao Paulo : Expressao Popular : UNESP. Postgraduate Program in Geography, 2009. 368 p.

SOTCHAVA, V. B. A theory for classifying geosystems of terrestrial life. IGEOG-USP. Sao Paulo. n. 14, 1978.

SOUZA, M. J. L. DE. The Territory: On space and power, autonomy and development. In: CASTRO, Iná Elias de, GOMES, Paulo Cezar da Costa,

TRICART, J. Ecodynamics. Rio de Janeiro: IBGE, 1977. 97p.

VIANA, S.B.R, A Fazenda Santa Cruz e a Política Real e Imperial em Relaçao ao Desenvolvimento Brasileiro (1790-1850). Master's dissertation. Niterói: Institute of Human Sciences and Philosophy, Fluminense Federal University. 1974. 288p.

WILSON, E. O. Biodiversity. Rio de Janeiro. Editora Nova Fronteira. 1992.

MIX
Papier aus verantwortungsvollen Quellen
Paper from responsible sources
FSC® C105338

Printed by Books on Demand GmbH, Norderstedt / Germany